Green-Schools Meeting the challenge of Climate Change

By

Anna Kavanagh

ISBN: 978-1-905451-77-7

A CIP catalogue for this book is available from the National Library.

This book was published in cooperation with Choice Publishing & Book Services Ltd, Ireland
Tel: 041 9841551 Email: info@choicepublishing.ie
www.choicepublishing.ie

Preface

The debate over climate change has ended. The scientific evidence is unequivocal. Global warming can no longer be denied because the evidence is all around us. Climate change is a moral concern and this presents a challenge for schools. It is of particular concern for religious run schools because of their Christian ethos. The vast majority of Irish children attend a Catholic school. All schools, but in particular those that are under the trusteeship of religious bodies are faced with the challenge of overcoming any denial that lingers about climate change and meeting head-on the challenge of doing something about it. World leader were slow to react at first to the mounting scientific evidence but are now involved in the growing momentum at a political level to reduce carbon emissions. The EU aims to take the moral high ground with a proposal to reduce greenhouse gas emissions in the EU by 363m tonnes, or 20%, by 2020. This effort will probably impact on all schools in the years ahead making it obligatory for them to reduce their carbon footprint. Even if it does not, there is a moral obligation on schools, especially those that are religious run to reduce carbon emissions. This book sets out to explore in detail the scientific evidence for climate change and the political response to it. It argues that schools, especially those that are religious run, are morally bound to face up to the challenge of global warming. Finally, it makes the case that the Green-Schools programme is a practical means of reducing carbon emissions and facing up to the challenge of climate change.

Foreword

Over the past decade a quiet revolution has been taking place in schools in Ireland and across the world. Schools, their students, teachers, parents and wider communities are successfully dealing with environmental and sustainability issues in their daily lives and achieving remarkable results. This quiet revolution is Green-Schools, internationally known as Eco-Schools. Pre-schools, Montessori, Primary, Secondary and Special Schools are all involved and are achieving reductions in the amount of waste they produce, energy they consume, water they use and are making their journey to and from schools more environmentally friendly, safe, healthy and fun. All this success is taking place in a background of a rather stark outlook and realisation of issues such as climate change, peak oil and dwindling world food and natural resources. Governments and administrations all over the world are struggling with these issues and although achieving some success they cannot match the results from Green-Schools. Irish schools involved in Green-Schools are diverting on average 65% of their waste away from landfill, saving between 25-33% on their energy usage, sometimes in schools with a very poor building fabric, saving millions of litres of high quality drinking water, (which we often take for granted in Ireland), and shifting away from tens of thousands of often unnecessary car journeys to and from schools and receiving the associated health and safety benefits.

However, Green-Schools is not just about the here and now, it is an investment in the further. Over the last decade hundreds of thousands of students have been involved in Green-Schools. Many students have gone from primary schools that are involved in the programme and continued and developed the programme in their secondary schools. Now our third level colleges and institutions are interested in starting the programme. This initiative and drive for the development of Green-Schools programme at third level coming from students that were involved in the programme at primary and secondary level. Furthermore, Green-Schools is not just confined to the school. Green-Schools has a big role to play in the wider community where positive 'pester-power' is making parents and other family members switch off standbys, buy loose at the supermarket, indulge in the occasional 'Telly-Free Tuesday' and take part in community clean ups and other community activities.

So, one would now ask what is Green-Schools, what's it all about? Green-Schools, (internationally known as Eco-Schools), is an international environmental and sustainable development education programme, environmental management system and award scheme that promotes and acknowledges long-term, whole school action for the environment. Green-Schools is actually one of the outputs from the Earth Summit in Rio in 1992. It was felt at that time that it was necessary to develop an environmental education and sustainable development education programme for schools. This challenge was taken up by the Foundation for Environmental Education (FEE), which is a federation and network of environmental education Non-Government Organisations (NGO's) from across the world. A number of education experts from FEE developed the programme from 1992-1995 and as they say the rest is history. Ireland, through it's FEE member An Taisce, joined the programme in 1997 and the programme has become one of the most successful education programmes of it's type in Ireland and is one of the leading countries within the FEE network of currently 42 countries with around 3,000 of the Ireland's 4,000 schools participating in the programme and the number constantly growing. The programme is based on a seven-step continual improvement, review and activity process and schools work through a number of themes such as litter & waste, energy, water and schools travel and cross-cutting themes such as climate change and future themes such as biodiversity and global & European citizenship. Schools that have successfully implemented and maintained the programme are awarded the very prestigious Green-Flag. There are currently around 1,450 Green-Flags flying in Ireland and the number continues to grow substantially every year. It is now impossible not to notice Green-Flags flying outside our schools and once ones spots the first they seem to appear to be almost everywhere.

The thing is though, Green-Schools is more that just a string of statistics of success. Green-Schools is also a great working example of partnership in action. Over the years a strong working relationship has developed between the schools, An Taisce and the Local Authorities to facilitate and develop the programme at a local level. Commercial and industrial sponsors, (some of whom have been with the programme since the very start), Government Departments and Agencies have come onboard over the years to fund the national development of programme and long may this continue. This funding has allowed the programme to develop new themes, expand and deal with the increased numbers of schools undertaking the programme.

However, as far as I am concerned Green-Schools is about great stories. It's about the voluntary and altruistic spirit among participants, particularly teachers, that cascades from the classroom into the homes and communities for the benefit of everyone. At the end of the day one has to remember that Green-Schools is a voluntary programme. How often are we reminded that volunteerism is a disappearing resource in our modern world? Green-Schools is about the school that wrote to a well known UK soap opera outlining what they were doing as regards Green-Schools and indicating that it would be a good idea to make some of the characters more environmentally friendly and furthermore receiving a lengthy reply from the scriptwriters that the school's letter had been discussed in a recent scriptwriter meeting and some of ideas would be taken onboard. It's about the Green-School students that made a presentation to their Local Town Councillors indicating the short falls of the towns litter management plan and how it could be improved; in essence embarrassing the adults into action. It's about the Green-School that after having unsuccessfully attempted worm composting decided upon getting a few hens for the schools. These dealt with the schools food scraps and also used shredded waste paper for bedding, (which mixed with the hen manure was a great activator for the composter). The hens had their summer and Christmas holidays like everybody else, but where housed in a chicken run in the schools which was made from old election posters! It's about the schools that discover while undertaking their water theme discovered that a school in Ethiopia needed funds to purchase water pumps to get water to their school. Funds were raised and the school in Ethiopia has it's water. It's all these stories and many others of resourcefulness, innovativeness and creativity. In essence Green-Schools is more than the sum of it's part's. In the challenging world today of climate change and increasing resource consumption we will require people to think as global citizens if irreversible damage is to be avoided. As the saying goes "Children make up 25% of the population but they are 100% of the future".

Dr. Michael John O Mahony

Green-Schools Development Officer

An Taisce.

For Maeve, Caoimhe and Paddy

We do not inherit the Earth from our Ancestors; we

borrow it from our Children

(Indian Proverb)

Acknowledgements

I'm eternally grateful to Bernadette Nestor my mentor and Anam Cara whose reassurance carried me through. I wish to thank Elizabeth Mc Ardle, IMU, Dalgan for her constant words of encouragement and support. A word of thanks to Dr John Sweeney, Head of the Irish Climate Analysis and Research Unit (ICARUS) at NUI Maynooth, who took time out of his busy schedule to give a public lecture in St. Joseph's Secondary School, Rochfortbridge, on the proposed effects of climate change. I'm thankful to Sonya Quinn, University of Limerick who has worked for a number of years to make it possible for St. Joseph's to become the first school in Ireland to calculate its ecological footprint. I wish to express my gratitude to Dr Michael John O' Mahony, An Taisce's Green-Schools Development Officer who has written the foreword and whose assistance was invaluable as I wrote this book. I am grateful to all my colleagues - staff and Green-School Committee members in St. Joseph's Secondary School, Rochfortbridge who helped me to implement the Green-Schools Programme, especially our Principal, Mr. Anthony Hartnett. Many thanks to my colleague Eugene Dunbar who came to my rescue every time I encountered an IT problem. Thanks to Mary Reilly, Department of Education and Science for providing statistics. I would like to say a special word of thanks to Dr. Sean Mc Donagh whose many books, especially *Climate Change: The Challenge to All of Us* provided me with the inspiration for this book. Sean, just before his departure for The United Nations Climate Change Conference in Bali, in December 2007, travelled to Rochfortbridge to give a lecture on climate change. Finally I would like to say thanks to Padraig whose reminiscences of the time he spent with the Native American Indians in Canada awakened my interest in ecology. To Nora and Aidan, my parents, who taught me how to live in a sustainable way, thank you.

'If you are thinking a year ahead, plant a seed.

If you are thinking a decade ahead, plant a tree.

If you are thinking a century ahead, educate the children.'

(Chinese proverb)

Table of contents

Introduction

Michel Jarraud, Secretary-General of the World Meteorological Organization (WMO) presented data to delegates at the Bali UN Summit on Climate Change in December 2007 showing that the decade 1998-2007 was the warmest on record.[1] He said that 20th century global average surface temperature has risen by 0.74°C. In 2007 winter and spring temperatures in parts of Europe were among the warmest ever recorded there. During the summer two extreme heat waves affected south-eastern Europe. Previous records were broken in June and July with daily maximum temperatures exceeding 40°C/104°F in some locations. Dozens of people died and fires destroyed thousands of hectares of land. A severe heat wave in the southern states of the USA during August resulted in 50 deaths attributed to excessive heat. In Australia, while conditions were not as severely dry as in 2006, long-term drought meant water resources remained extremely low in many areas. Below average rainfall over the densely populated and agricultural regions resulted in significant crop and stock losses, as well as water restrictions in most major cities.

The WMO also recorded severe flooding in many parts of the world throughout 2007.[2] Mozambique experienced its worst flooding in six years, killing dozens, destroying thousands of homes and flooding 80,000 hectares of crops in the Zambezi valley. In Sudan, torrential rains caused flash floods in many areas in affecting over 410,000 people, including 200,000 left homeless. Monsoons triggered widespread flash floods in West Africa, Central Africa and parts of the Greater Horn of Africa. Some 1.5 million people were affected and hundreds of thousands of homes destroyed. In Bolivia, flooding affected nearly 200,000 people and 70,000 hectares of cropland. Uruguay was hit by its worst flooding since 1959, with heavy rain causing floods that affected more than 110,000 people and severely damaged crops and buildings. Massive

[1]Michel Jarraud, Statement at the high-level segment of the thirteenth session of the Conference of the Parties to the United Nations Framework Convention on Climate Change, Bali, Indonesia, 13 December 2007, http://www.wmo.ch/pages/cop13/statement_en.html

[2] Ibid

flooding in Mexico destroyed the homes of half a million people and seriously affected the country's oil industry. In Indonesia floods covered half of the city of Jakarta by up to 3.7 metres of water. Heavy rains ravaged areas across southern China, with flooding and landslides affecting over 13.5 million people and killing more than 120. Monsoon-related extreme rainfall events caused the worst flooding in years in parts of South Asia. About 25 million people were affected in the region, especially in India, Pakistan, Bangladesh and Nepal. Thousands lost their lives. England and Wales recorded its wettest May-July period since records began in 1766, receiving 406 mm of rain compared to the previous record of 349 mm in 1789. Extensive flooding in the region killed nine and caused more than US$6 billion in damages.

A number of storms and hurricanes were also recorded. A powerful storm system, Kyrill, affected much of northern Europe during 17-18 January 2007 with torrential rains and winds gusting up to 170km/h. There were 47 deaths across the region, with disruptions in electricity supply affecting tens of thousands. During the 2007 Atlantic Hurricane season, 14 named storms occurred, compared to the annual average of 12. For the first time since 1886, two category 5 hurricanes (Dean and Felix) made landfall in the same season. In June, cyclone Gonu made landfall in Oman, affecting more than 20,000 people and killing 50, before reaching the Islamic Republic of Iran. There is no record of a tropical cyclone hitting Iran since 1945.On 15 November, tropical cyclone Sidr made landfall in Bangladesh, generating winds of up to 240 km/h. More than 8.5 million people were affected and over 3,000 died. Nearly 1.5 million houses were damaged or destroyed.

Another remarkable global climatic event recorded in 2007 by the WMO was the first recorded disappearance of ice across parts of the Arctic that opened the Canadian Northwest Passage for about five weeks starting in August. Nearly 100 voyages in normally ice-blocked waters sailed without the threat of ice. Melting ice led to a rise in sea levels of 1.7 mm which is substantially above the annual average for the 20th century. Mr. Jarraud noted that scientists had recently forecasted that the ice could melt entirely during summers by 2013. In January 2008, provisional figures from Met Eireann, revealed that 2007 was the warmest year in Ireland since records began in the 1880s. Average temperatures for the country as a whole were at least 1 degree higher than normal.[3]

In May 2008, twelve days after Cyclone Nargis struck Burma, John Holmes, the UN humanitarian affairs chief, told reporters that between 1.6 million and 2.5 million people were without shelter following the storm which struck at the beginning of the month. The Red Cross said it believed up to 130,000 people may have been killed by the cyclone. Official figures are significantly lower, reporting 34,273 killed and 27,838 missing.[4]

In June 2008 Environment Minister John Gormley unveiled a scientific report that plots the potentially catastrophic effects of global warming on the Irish environment. The report, *Ireland in a Warmer World: Scientific Predictions of the Irish Climate in the 21st Century*, states that our climate will continue to warm up, particularly in the summer and autumn seasons. The study was conducted by the Community Climate Change Consortium for Ireland, also known as the C4I Project, and was based on extensive studies carried out by Met Éireann and UCD. It describes Ireland's changing climate and is based on a comprehensive series of computer simulations. It states that the changing climate could have far-reaching consequences for agriculture and domestic water supplies. The south and east of the country have been identified as areas that will experience the most dramatic effects of global warming with temperature increases of three to four degrees by the end of the century. The projections in the latest analysis show that rainfall is expected to increase in winter by about 15% and summer predictions range from no change to a 20% decrease. It shows that Ireland's average temperature has increased by about 0.7C over the past 100 years and the rate of increase has been higher in the past couple of decades.[5]
Also in June 2008, Met Éireann says that the previous month was the warmest May on record in most places. Mean air temperatures

---

[3] Ronan Mc Greevy, *Warmest year since the 1880s*, Irish Times 31/12/07, page 9
[4]Ian MacKinnon, *2.5m Burmese may be homeless*, The Guardian. Wednesday May 14, 2008
[5]' Minister Gormley launches a new report *Ireland in a Warmer World: Scientific Predictions of the Irish Climate in the Twenty-First Century.*'
http://www.environ.ie/en/Environment/MetEireann/MainBody,17545,en.htm
10/06/08

for May were above normal everywhere and were more than three degrees higher than normal in western areas.[6]

Chapter one will demonstrate that this extreme weather in is not a series of one-off events but part of a well established pattern of climate change that is happening at a pace much faster than at any other time in the earth's history. It will show that the majority of the world's scientists agree that human activity - mainly greenhouse gas emissions from the burning of fossil fuels are responsible and the danger is we are moving outside natural ranges and towards 'tipping points' beyond which very large-scale consequences may be irreversible. Poor and vulnerable communities least responsible for causing the problem will be hit first and hit the hardest.

Chapter two will demonstrate that present day warming has already caused a range of negative effects and will continue to create far more unless halted. A day hardly passes without some report in the media of a weather related incident. As I write in June 2008, China's Civil Affairs Ministry reports that 1.66 million people have been evacuated from the hardest-hit areas following flooding in nine provinces in the south.[7] Meanwhile in the USA, flooding in Iowa has devastated large areas of farmland and sent corn prices to a record high of $8 a bushel. Iowa is the biggest corn producer in the US.[8] Displacement of people and soaring food prices are just two of a myriad of effects of climate change which are examined in detail.

Chapter three traces the international political response to climate change from the Protocol agreed in Kyoto in December 1997 to the recent agreement signed in Bali in December 2007. Meanwhile in Ireland the government has drawn up the *National Climate Change Strategy,* a blueprint to help us meet our commitments under international law to reduce carbon emissions. I argue that our political representatives are not doing enough at present to combat

---

[6] Collins Dan, *Global warming fears for south and east, Irish Examiner*, 10 June 2008, page 1

[7] BBC, *More floods threaten south China*, http://news.bbc.co.uk/2/hi/asia-pacific/7459628.stm, Tuesday, 17 June 2008.

[8] BBC, *Midwest braced for more flooding*, http://news.bbc.co.uk/2/hi/americas/7458220.stm, Tuesday, 17 June 2008.

climate change and maintain that if we are really serious about the issue, then a good place to start is in the school.
Chapter four argues that facing up to the challenge of climate change is an inescapable moral responsibility for theologians. However Catholic Church leaders are not doing enough at present to stress the urgency of this issue. This has major implications for schools. More than 3000 of the 3,200 primary schools in the Republic of Ireland are controlled by the Catholic Church. Over half of the 732 second level schools are religious run. I will show that there is a problem for Catholic schools because the Catholic Church has not yet published a pastoral letter on climate change nor have they issued a statement stressing the urgency of the matter for those concerned with education. This has serious repercussions for Irish children because the majority of them attend a Catholic school.

Chapter five proposes that the Green-Schools programme is a long-established programme which is an effective method of meeting the challenge of climate change. However, it is my experience that not enough is being done at present to promote and support it. In September 2005 I signed up for the MA (Ecology and Theology) programme in Dalgan Park, Navan and also volunteered to become the Green-Schools Co-ordinator in St Joseph's, Convent of Mercy Secondary School, Rochfortbridge where I work as a teacher. For a long time I saw no connection between the two. In Dalgan I was learning how man is destroying life on the planet and the moral obligation for Christians to do something about this. I was particularly moved by Sean Mc Donagh's book *Climate Change-the Challenge to All of Us*. Slowly it dawned on me that my effort to develop the Green-School programme in our school was a practical means of meeting the challenge of climate change because it was challenging us to reduce our ecological footprint. However my efforts were fraught with difficulties because it is a voluntary programme and lacks proper support from the various agencies governing the school. I came to believe that Green-Schools is a practical means of meeting the challenge of climate change at a localised level and for this reason I argue that it must be put on a statutory footing and given the necessary funding to enable it to meet the challenge of climate change.

"Making a difference in the world should not depend on governments but on people themselves. If you want to see change, be the change you want."

Dr. Tony Humphreys.

Chapter 1

Introduction

In 1824 Jean Baptiste Joseph Fourier discovered that greenhouse gas traps heat radiated from the Earth's surface after it has absorbed energy from the sun.[1] In 1859 the British physicist, John Tyndall proposed that ice ages were caused by a decrease in the amount of atmospheric carbon dioxide.[2] In 1896 Svente Arrhenius showed that doubling the carbon dioxide content of the air would slowly but surely raise global temperatures by 5-6C.[3] It was not until 1938 that there were further advances in the climate change theory when Guy S Callender discovered that the burning of fossil fuels caused a 10% increase in atmospheric carbon dioxide.[4] However, scientists still believed that most of the carbon dioxide produced by humans dissolved safely in the oceans. This hypothesis was rejected in 1957 by Hans Suess who discovered an intricate system which prevents sea water from absorbing too much atmospheric carbon dioxide.[5] Scientists finally started taking the notion that humans could contribute to global warming seriously. Evidence to test if global temperatures were increasing alongside greenhouse gas emissions was now being collected and used to construct mathematical models to predict future climates. To the fore was Charles Keeling, who in 1958 began taking long-term measurements of atmospheric

---

[1] Sean Mc Donagh, *Climate Change-The Challenge to All of Us*, Columba Press, 2006, page 14.
[2] Ibid, page 15
[3] Ibid, page 15
[4] Callendar, G.S. (1938). *The Artificial Production of Carbon Dioxide and Its Influence on Climate*. Royal Meteorological Society Quarterly, Vol. 64, pages 223-40.

[5] Hans E. Suess, 'Natural Radiocarbon and the Rate of Exchange of Carbon Dioxide between the Atmosphere and the Sea.' In *Nuclear Processes in Geologic Settings*, edited by National Research Council Committee on Nuclear Science, National Academy of Sciences, Washington, D. C.,(1953), pp. 52-56.
[6] Sean Mc Donagh, *Climate Change-The Challenge to All of Us,* page 16

carbon dioxide at the Mauna Loa observatory in Hawaii.[6] His figures revealed an indisputable annual increase. They showed a 35% increase in CO_2 levels compared to pre-industrial levels and also that they were higher than at any time in the previous 700,000 years. In addition he showed that atmospheric temperatures had increased by 0.5-0.6C over the previous 150 years.

From the 1960's evidence of global warming continued to accumulate. In 1999 Michael Mann published a detailed analysis of global average temperature over the previous millennium.[7] His discovery became known as the 'hockey stick graph' because when it was plotted onto a chart, that is what it resembled. The graph is relatively flat from the period 1000 to 1900 AD, demonstrating that temperatures were comparatively stable for this period of time. The flat part forms the stick's 'shaft.' After 1900, however, temperatures appear to shoot up, forming the hockey stick's 'blade'. Given that there is a sharp rise in temperature since the industrial revolution it is assumed that climate change is caused by human activities. This graph featured in the Intergovernmental Panel on Climate Change (IPCC) Third Assessment Report. Data coming from satellites, weather stations around the world and thousands of research papers are now able to give a clearer picture of what is happening. It's the function of the Intergovernmental Panel on Climate Change (IPCC) to correlate this data and reach a verdict.

Intergovernmental Panel on Climate Change (IPCC)

The IPCC is the science authority for the United Nations Framework Convention on Climate Change (UNFCCC). Its task is to assess the risk of climate change caused by human activity. The World Meteorological Organization (WMO) and the United Nations Environment Programme (UNEP) set up the panel in 1988. It does not carry out research, nor does it monitor climate. Its function is to publish special reports on topics relevant to the implementation of the UN Framework Convention on Climate Change (UNFCCC). The IPCC bases its assessment on reviewing published scientific literature on climate change. The panel is made

---

[7] Mann, Michael E., et al. (1999). 'Northern Hemisphere Temperatures During the Past Millennium: Inferences, Uncertainties, and Limitations.' *Geophysical Research Letters 26*: pp. 759-62.

up of 450 authors and 800 contributors, while 2,500 scientific experts take part in the hearings. Governments take part in the review of the reports.

Its first assessment report was published in 1990 and became one of the basic documents at the UN Conference on Environment and Climate in 1992. It stated that emissions resulting from human activities were substantially increasing the atmospheric concentrations of CO_2, methane, CFCs and nitrous oxide.[8] It predicted that these increases would have a greenhouse effect. It also maintained that CO_2 has been responsible for over half the greenhouse effect. The report also envisages an increase of global mean temperature during the 21st century of about 0.3 Degrees Celsius per decade. To put this in context, this is a greater increase than that seen over the past 10,000 years.

The IPCC Second Assessment Report on Climate Change was published in 1995. It found that greenhouse gas concentrations were continuing to increase.[9] It also stated that their predictions on air temperature increases had not changed significantly since the 1990 report. However this report was more forthright in stating that the outlook for planet earth was not good: It stated that interference with the climate system from greenhouse gas emissions would 'grow in magnitude and the likelihood of adverse impacts from climate change that could be judged dangerous will become greater.' The IPCC Second Assessment Report became the foundation for the Kyoto Agreement of 1997, which has played such a significant part in the international debate on global warming.

The IPCC Third Assessment Report on Climate Change was published in 2001. It found that there was a global average surface temperature increase of 0.6 °C over the course of the 20th century with the 1990s being the warmest decade and 1998 the warmest

---

[8] JT Houghton, GJ Jenkins and JJ Ephraums (Eds), *Scientific Assessment of Climate change – Report of Working Group 1,* Cambridge University Press, 1990
[9] IPCC, *Second Assessment Synthesis of Scientific-Technical Information relevant to interpreting Article 2 of the UN Framework Convention on Climate Change*, 1995, page 2, http://www.ipcc.ch/pdf/climate-changes-1995/2nd-assessment-synthesis.pdf Retrieved 23/09/2007

year since records began.[10] It stated that there was new and stronger evidence to show that most of the warming observed over the previous fifty years was attributable to human activities.[11] It predicted that if current trends continued, average surface temperature would increase by 1.4 to 5.8 Celsius degrees over the period 1990 to 2100. They warned that this would lead to the extinction of some endangered species, water shortages and severe social and economic effects arising from sea-level rises and storms. It also warned that the impacts of climate change would fall disproportionately upon developing countries.[12]

The Fourth IPCC Assessment Report, the most detailed to date, was published in February 2007. It includes the contributions of more than 1,200 authors and 2,500 scientific expert reviewers from more than 130 countries. The report finds that it is very likely' that emissions of heat-trapping gases from human activities have caused 'most of the observed increase in globally averaged temperatures since the mid-20th century.'[13] It presents scientific evidence to demonstrate that human activities are the major cause of recent climate change. The report claims that it is 'unequivocal' that Earth's climate is warming as is now evident from observations of increases in global average air and ocean temperatures, widespread melting of snow and ice and rising global mean sea level. The report also confirms that the current atmospheric concentration of carbon dioxide and methane, two important heat-trapping gases, exceeds by far the natural range over the last 650,000 years. Since the beginning of the industrial era concentrations of both gases have increased at a rate that is very likely to have been unprecedented in more than 10,000 years. The report points out that eleven of the last 12 years rank among the 12 hottest years on record (since 1850, when sufficient worldwide temperature measurements began). It says that the intensity of tropical cyclones (hurricanes) in the North Atlantic have increased over the past 30 years and storms with heavy precipitation have increased in

---

[10] IPCC, 'Technical Summary', *Climate Change 2001: Synthesis Report*, pp.26-28 http://www.ipcc.ch/ipccreports/tar/vol4/english/pdf/wg1ts.pdf Retrieved 23/09/2007

[11] Ibid, pp 35-39

[12] IPCC, 'Summary for Policy Makers', *Climate Change 2001*, pp 8-14, http://www.ipcc.ch/ipccreports/tar/vol4/english/pdf/spm.pdf Retrieved 23/09/2007

[13] IPCC, Fourth Assessment Report, Synthesis Report http://www.ipcc.ch/pdf/assessment-report/ar4/syr/ar4_syr_topic1.pdf

frequency over most land areas. Meanwhile, the Sahel (the boundary zone between the Sahara desert and more fertile regions of Africa to the south), the Mediterranean, southern Africa, and parts of southern Asia have become drier, adding stress to water resources in these regions. Droughts have become longer and more intense, and have affected larger areas since the 1970s, especially in the tropics and subtropics. Since 1900 the Northern Hemisphere has lost seven percent of the maximum area covered by seasonally frozen ground. Mountain glaciers and snow cover have declined worldwide. Satellite data since 1978 show that the extent of Arctic sea ice during the summer has shrunk by more than 20 percent. Since 1961, the world's oceans have been absorbing more than 80 percent of the heat added to the climate, causing ocean water to expand and contributing to rising sea levels. Between 1993 and 2003 ocean expansion was the largest contributor to sea level rise. In addition, melting glaciers and losses from the Greenland and Antarctic ice sheets have also contributed to recent sea level rise.

This latest report from the IPCC predicts that global warming will continue giving rise to more intense tropical cyclones in some places, extreme heat waves in others and heavy precipitation resulting in flooding in other parts. Sea ice is projected to shrink in both the Arctic and Antarctic under all model simulations. Some projections show that by the latter part of the century, late-summer Arctic sea ice will disappear almost entirely. Increasing atmospheric carbon dioxide concentrations will lead to increasing acidification of the ocean, with negative repercussions for all shell-forming species and their ecosystems. They predict that global average sea level will rise between 7 and 23 inches (0.18 and 0.59 meters) above the 1980–1999 average. The report concludes that even if we act today to reduce our emissions from cars, power plants, land use, and other sources, we will see some degree of continued warming because past emissions will stay in the atmosphere for decades or more. If we take no action to reduce emissions, the IPCC concludes that there will be twice as much warming over the next two decades than if we had stabilized heat-trapping gases and other climate relevant pollutants in the atmosphere at their year 2000 levels.

Critics of the report were silenced when the Nobel Peace Prize was awarded in October 2007 to the Intergovernmental Panel on

Climate Change and Al Gore 'for their efforts to build up and disseminate greater knowledge about man-made climate change, and to lay the foundations for the measures that are needed to counteract such change.'[14] The Fourth IPCC Assessment Report ends the debate on climate change and firmly establishes them as the leading scientific authority on climate change.

Critics of the IPCC

A month after the publication of the Fourth IPCC Report Richard Lindzen appeared in *The Great Global Warming Swindle*, a Channel Four documentary broadcast in the UK in March, 2007. Dr. Lindzen is an atmospheric physicist and the Alfred P. Sloan Professor of Meteorology at the Massachusetts Institute of Technology. He was the lead author of Chapter Seven of the IPCC Third Assessment Report. In the documentary he was put forward as the most eminent scientific critic of climate change. The documentary sent out a clear signal that there are noteworthy scientists unconvinced by the climate change argument. However when one delves a little deeper into Lindzen's contributions to date on the climate change debate, one finds that in an article he wrote for the *The Hill Times* in 2004, he describes the Third IPCC report as 'an admirable description of research activities in climate science.'[15] However he is highly critical of the Summary for Policy Makers (SPN) which in his opinion 'has a strong tendency to disguise uncertainty, and conjures up some scary scenarios for which there is no evidence.' He points out that the SPM claims that greenhouse gases are accumulating in Earth's atmosphere as a result of human activities, causing surface air temperatures and subsurface ocean temperatures to rise mainly during the previous twenty years. However the full report states that twenty years is too short a period for estimating long-term trends, a crucial point that the SPM fails to mention. Of major significance is the fact that he accepts that warming has occurred and calculates that the global

---

[14] Nobel Foundation, *The Nobel Peace Prize 2007*, http://nobelprize.org/nobel_prizes/peace/laureates/2007

[15] Richard Lindzen, *Canadian Reactions To Sir David King*, The Hill Times Ottawa ,Feb 23 - March 1, 2004 http://meteo.lcd.lu/globalwarming/Lindzen/canadian_reactions_to_sir_david_king.html

[16] Michael Nortcott, *A Moral Climate: The Ethics of Global Warming*, Darton,Longman & Todd Ltd., 2007 page 280

mean temperature is about 0.6 degrees Celsius higher than it was a century ago and that atmospheric levels of carbon dioxide have risen about 30 percent over the past two centuries. Furthermore, he states that the carbon dioxide 'increase is likely to warm the earth.' What is clear here and also in the documentary is the fact that he is not denying that climate change is occurring but he has a huge problem with the way in which some people are making alarmist statements. He wrote this article in response to a statement made by Sir David King, Chief Scientific Adviser to the UK Government who compared climate change to global terrorism. 'Equating climate change with global terrorism, as both the environment minister and Dr. King have done recently, is precisely the sort of statements that all concerned, thinking citizens should condemn.' he wrote.

Theologian Michael Northcott who has written on the ethics of climate change also denounces Dr. King. He says that 'it would then be a tragic irony if politicians or scientists were to resort to the politics of fear to promote the need to tackle global warming. And yet some, as we have seen, have already resorted tol this tactic, including Al Gore, Sir David King, and a number of other prominent voices in the climate change debate.'[16] In his opinion 'Global Warming is in one sense, of course, a greater threat to human security than terrorism. But the politics of fear has no more resolved the causes of if terrorism than it will resolve the causes of global warming.' There is merit in what Lindzen has to say but to categorise him as one who is in denial about climate change is to distort the facts.

Support for the Scientific Argument

The scientific understanding of climate change dates back to Fourier in the 1820's and has more or less concluded with the publication of the Fourth IPCC Assessment Report in February 2007.Two years prior to the publication of this report, Naomi Oreskes, a scientist at the University of California, analysed 923 articles in scientific journals written between 1993 and 2003. She published the result of her findings in *Science* in December 2004.[17] She concluded that climate change is taking place and that not a single article reviewed disputed the fact that human activity was causing a rise in global temperatures.

Mc Donagh, referring to the Pontifical Council for Justice and Peace two-day seminar on climate change held in May 2007, attended by over 80 people, says that there were excellent presentations from credible scientists, whose writings have been peer reviewed.[18] However there were at least four participants who either deny climate change or believe it is a good thing. One of these was Professor Craig Idso, adjunct professor at the Office of Climatology at Arizona State University (ACU). He is chairman of the *Centre for the Study of Carbon Dioxide and Global Change,* an institute dedicated to denying climate change. Together with his father Sherwood and his brother Keith, he co-authored a report entitled '*Enhanced or Impaired? Human Health in a CO2-enriched world,*' which argues that global warming and an increase in atmospheric CO2 would be beneficial to humanity. Mc Donagh reports that earlier in the year the *Union of Concerned Scientists* released a report documenting how ExxonMobile had given millions of dollars to individuals and groups who raise doubts about climate change. The *Center for the Study of Carbon Dioxide and Global Change* is listed as being an ExxonMobil-funded group. Meanwhile, he writes that in September 2006, Bob Ward, Senior Manager for Policy Communication for the Royal Society, Britain's premier scientific body, wrote to ExxonMobile demanding that the company stop funding groups that have, 'misrepresented the science of climate change by outright denial of evidence'. Northcott says that 'the claim that the science of global warming is unreliable has for many years been sustained by the public relations departments of the world's largest oil companies, who have clear interests in denying global warming. Their profit margins might be seriously affected if governments drastically reduce fossil-fuel dependence. They are the largest producers of

---

[17] Oreskes Naomi, *Beyond the Ivory Tower : The Scientific Consensus on Climate Change*, Science 3 December 2004: Vol. 306. No. 5702, p. 1686 www.sciencemag.org/cgi/content/full/306/5702/1686 Retrieved 15/08/2007

[18] Sean McDonagh, *Climate change, one of the most serious moral issues of our times,* J/P Alert, July/August 2007, http://www.cmsm.org/CMSM_Alert/JulAug07/ Retrieved 10/10/2007

fossil fuels and they may be held legally liable in an international court at some point in the future for compensating victims of climate change.'[19] He explains that 'the world's largest company, ExxonMobil, is the principal source of funding for the scientists and lobby groups who continue to feed global media with the claim that there are two sides to the global warming debate, and that a balanced presentation of the evidence ought to represent the believers and the nay-sayers. The company funds groups which suggest that climate change science is not sound science and insist that considerable uncertainties remain in the outcomes predicted by climate scientists.'

Many leading scientists who doubted climate change over the years are slowly but surely being convinced by the overwhelming body of evidence being presented by the IPCC and other bodies such as the World Wildlife Fund. Sir David Attenborough has been a leading sceptic of climate change over the years. However in recent years, like others, he has had a change of mind. Writing in the London *Independent* in 2004 he said: 'I have waited until the proof was conclusive that it was humanity changing the climate. The thing that really convinced me was the graphs connecting the increase of carbon dioxide in the environment and the rise in temperature, with the growth of human population and industrialisation. The coincidence of the curves made it perfectly clear we have left the period of natural climatic oscillation behind and have begun on a steep curve, in terms of temperature rise, beyond anything in terms of increases that we have seen over many thousands of years.'[20] Speaking to the Commons environment committee in December 2006, he said there was no question that global warming would worsen. 'What we can do is make the situation deteriorate less than it's going to.'[21] He went on to tell them that if no action was taken on emissions, there was more than a 75% chance of global temperatures rising between two and three degrees Celsius over the next 50 years. Sir David warned that 'it's

---

[19] Michael Northcott, *A Moral Climate: The Ethics of Global Warming,* page 270

[20] David Attenborough, *Climate Change is the Major Problem Facing the World,* Independent/UK, May 24, 2006, http://environment.independent.co.uk/article570935.ece , retrieved 21/09/2007

[21] BBC, *Attenborough urges 'moral change,* http://news.bbc.co.uk/2/hi/uk_news/politics/6176847.stm 13/12/06 retrieved 14/11/2007

morally wrong to waste energy because we are putting at risk our grandchildren.'

Earlier in the year he presented a two-part television programme *Are We Changing Planet Earth* that explored how climate change is altering the planet. Speaking ahead of the broadcast he said that 'If we do care about our grandchildren then we have to do something, and we have to demand that our governments do something.'[22]

What if the scientists are wrong?

There are many who argue that scientists have been proved wrong in the past and that there is a possibility that they may be proved wrong once again on the issue of climate change. Northcott says that 'even if the deniers were right - which is impossible to credit on rational grounds - the core argument …. is that the fossil-fuelled global economy is dangerous to planet earth and to human life, not just because it is warming the climate of the earth but because it is deeply destructive of the diversity and welfare of the ecosystems and human communities from which surplus value is extracted and traded across highways, oceans and jetstreams.'[23] He explains that new practices encouraged by the recognition of global warming such as turning off lights, turning down the heating, cycling or walking instead of driving, holidaying nearer to home, buying local food, shopping less and conversing more, addressing the causes of fuel poverty locally and internationally are good because they are intrinsically right, not just because they have the consequence of reducing carbon emissions. Such actions correct modern thoughtlessness. They sustain the moral claim that it is wrong to live in a civilisation that depends upon the systematic enslavement of peoples and ecosystems to the high resource requirements of a corporately-governed consumer economy.

He explains the case made by the French philosopher Blaise Pascal who developed a famous response to arguments for and

---

[22] BBC, *Attenborough: Climate is changing,* http://news.bbc.co.uk/2/hi/science/nature/5012266.stm 24 May 2006, retrieved 14/11/2007

[23] Michael Northcott, *A Moral Climate: The Ethics of Global Warming,* page 273

against the existence of God.[24] He suggested that if, against the claims of atheists God does exist and atheists live morally heedless lives, then they risk judgment and even eternal damnation. However, if they take a wager and live as though God exists, they have little to lose since the kind of life that Christians traditionally commend 'fosters companionship and civic responsibility and a concern for the common good.' He goes on to explain how Stephen Heller proposes the same 'Pascalian wager' in relation to global warming: 'action to stem climate change would be prudent even if certain knowledge that it is happening, or about the severity of its effects, is not available or not believed. If global warming is humanly caused, then these actions will turn out to have been essential for human survival and the health of the biosphere. In the unlikely event that it is not, then these good actions promote other goods — ecological responsibility, global justice, care for species - which are also morally right.'[25]

He concludes: 'Actions which will have the effect of mitigating climate change are also actions which reaffirm the embodied relationship between inner desire and the outer world of what Christians call Creation. For this reason, such actions are intrinsically good, and will promote flourishing even if, as a minority of dissenters suggest, greenhouse gases are not the primary driver of global warming.'

Conclusion

The report of the Intergovernmental Panel on Climate Change, the largest and most rigorously peer-reviewed scientific consensus in history, advises that climate change is real, it is caused by human activity and it is threatening the planet in ways we can only begin to imagine. Sir John Houghton writing about his term as Co Chairman of the scientific assessment for the IPCC 1988-2002, says that several thousand scientists drawn from many countries were involved as contributors and reviewers in these assessments. Their task was 'honestly and objectively to distinguish what is reasonably well known and understood from those areas with large

---

[24] Ibid, page 274
[25] Ibid, page 274

uncertainty and to present balanced scientific conclusions to the world's policymakers. No assessment on any other scientific topic has been so thoroughly researched and reviewed.'[26] He explains that in June 2005, just before the G8 Summit in Scotland, the Academies of Science of the world's eleven most important countries (the G8 plus, India, China and Brazil) issued a statement endorsing the conclusions of the IPCC and urged world governments to take urgent action to address climate change. 'The world's top scientists could not have spoken more strongly.' Yet there is an ongoing public debate about how to respond to this challenge and about whether or not climate change is even a scientific certainty. 'Unfortunately, there are strong vested interests that have spent tens of millions of dollars on spreading misinformation about the climate change issue. First they tried to deny the existence of any scientific evidence for rapid climate change due to human activities. More recently they have largely accepted the fact of anthropogenic climate change but argue that its impacts will not be great, that we can 'wait and see' and in any case we can always 'fix' the problem if it turns out to be substantial. The scientific evidence cannot support such arguments.'[27]

Northcott warns that 'without a mass die-off of human beings, can the earth be healed from anthropogenic climate change, or is it already too late? If it is, will future generations forgive present industrial consumers?' He imagines what they might say to us should they meet us in heaven: 'But you knew the science was coming right, you knew the predictions were real, and yet you carried on burning this stuff and wrecking the earth. Why? Why did you not change before it was too late?' One of the reasons we would need to give in response is that we were 'in denial about our addictions.'[28] The first major challenge for everyone in the field of education is to accept the scientific argument for climate change.

---

[26] John Houghton, *Global Warming, Climate Change and Sustainability-Challenge to Scientists, Policy Makers and Christians,* JRI Briefing Paper NO 14, 2006,page 7, http://www.jri.org.uk/brief/Briefing14_print.pdf
[27] Ibid, page 7
[28] Michael Northcott, *A Moral Climate: The Ethics of Global Warming,* page 269

‘All across the world, in every kind of environment and region known to man, increasingly dangerous weather patterns and devastating storms are abruptly putting an end to the long-running debate over whether or not climate change is real. Not only is it real, it's here, and its effects are giving rise to a frighteningly new global phenomenon: the man-made natural disaster.’

Barack Obama

Chapter 2

Introduction

An assessment by the UN of the state of the world's environment published in October 2007 painted its bleakest picture yet of our planet's well-being. The report by the United Nations Environment Program[1] warns that humanity's future is at risk unless urgent action is taken. It demonstrates how the previous twenty years have resulted in almost every index of the planet's health deteriorating while at the same time, personal wealth in the richest countries has grown by a third. It warns that the vital natural resources, which support life on Earth, have suffered significantly since their first such report, published in 1987. However, the depletion of the world's *natural capital* has coincided with unprecedented economic gains for developed nations, which, for many people, have masked the growing crisis. It says humanity could be at risk if nothing is done to address climate change and the mass extinction of animals and plants. Energy consumption has risen sharply since the publication of the first report while the concentration of carbon dioxide is a third higher than it was then. It warns of catastrophic consequences if the issue of climate change is not addressed immediately. This chapter sets out to examine some of these consequences.

Natural Disasters

According to the *Centre for Research on the Epidemiology of Disasters*, a UN-backed research group, more than 16,500 people were killed worldwide and $62.5 billion (€42.7 billion) damage was caused due to climate related natural disasters in 2007.[2] Almost 200 million people were affected. Eight of the worst disasters last year struck Asia. *Cyclone Sidr* in Bangladesh in November claimed the highest toll of 4,234 lives. Europe's winter storm *Kyrill* caused

---

[1] UNEP, *Global Environment Outlook 4: Our Common Future,*25/10/2007 http://www.unep.org/geo/geo4/media/media_briefs/Media_Briefs_GEO-4%20Global.pdf. Retrieved 11/12/2007

[2] Bulletin Page, *Asia hit hardest by natural disasters in 2007*, Irish Times 19/01/2008 Page 24

$10 billion in damage. Closer to home, summer floods in Britain caused $8 billion in damage.Centre director Debarati Guha-Sapir warned that climate change caused by human emissions of greenhouse gases will bring extreme weather, including more heatwaves, droughts, floods and rising seas in coming years.

Agflation

Climate change is resulting in 'agflation' a term that looks set to become part of everyday vocabulary. An *Irish Times* report in June 2007 cites research conducted by William Gavin, Vice President of the US Federal Reserve, which shows that a core measure of inflation which includes food but not energy has proved to be a far more reliable indicator of the future trend in consumer prices than the traditional measure that excludes both.[3] It says that a convergence of factors suggests that food prices could easily double in the next seven years giving rise to *agflation*. This is because growth in global demand for agricultural commodities is accelerating particularly in China and India where the combined population should increase by almost 120 million over the next five years to more than 2.5 billion. The acceleration in demand for food comes at a time when supply is struggling to keep pace. Excessive withdrawal from surface water and from underground aquifers partly as a result of climate change has seriously depleted water resources in both countries. In China's Qinhai province, more than 2,000 lakes have vanished in the past 20 years, while the water table in the Punjab, India's bread basket, is falling by almost one metre every year. This is a direct result of climate change, higher temperatures leads to increased evaporation and increased drought results in aquifers being depleted. It notes that when agriculture and industry compete for scarce water, it is the farming community that inevitably loses out given that value-added output per ton of water is a mere fraction of the comparable figure for manufacturing. Water scarcity is already showing up in China's grain harvest, which has dropped by more than 25 per cent over the past decade. Falling grain supplies in China and India are being aggravated by energy policy in the US, which has seen an increasing proportion of the country's corn crop being used for ethanol production in a bid to

---

[3] Charlie Fell, *Seeds of Change, food prices could easily double in the next seven years*. Page 9, Irish Times,8th June 2007.

reduce carbon emissions. The surge in energy demand has led to a sharp rise in corn prices and food price inflation. Jean Ziegler, A UN expert has called the growing practice of turning crops into biofuel 'a crime against humanity' because it has created food shortages and sent food prices soaring, leaving millions of poor people hungry.[4] He said that 'the effect of transforming hundreds and hundreds of thousands of tons of maize, of wheat, of beans, of palm oil, into agricultural fuel is absolutely catastrophic for the hungry people.' He explained that the world price of wheat doubled in one year and the price of maize quadrupled, leaving poor countries, especially in Africa, unable to pay for the imported food they need to feed their people — and the poor people in those countries unable to pay the soaring prices for food.

In November 2007 John Vidal, environment editor of The Guardian Newspaper wrote about empty shelves in Caracas, food riots in West Bengal and in Mexico.[5] He said that warnings had been issued of impending hunger in Jamaica, Nepal, the Philippines and sub-Saharan Africa because of soaring prices for basic foods. This, according to him is beginning to lead to political instability, with governments being forced to step in to artificially control the cost of bread, maize, rice and dairy products. He noted that according to the UN Food and Agricultural Organization record world prices for most staple foods had led to 18% food price inflation in China, 13% in Indonesia and Pakistan, and 10% or more in Latin America, Russia and India. Meanwhile wheat had doubled in price, maize had increased by 50% and rice by 20% in the previous year resulting in India, Yemen, Mexico, Burkina Faso and several other countries experiencing or being close to food riots. Closer to home Italians organized a one-day boycott of pasta in protest at rising prices. The price rises are a result of farmers switching from cereals to grow biofuel crops to reduce carbon emissions and also because of reduced crop yields as a result of extreme weather conditions caused by climate change. The Australian government, for example had recently announced that drought had slashed predictions of

---

[4] Edith M. Lederer, *Production of biofuels is a crime',* London Independent http://environment.independent.co.uk/green_living/article3101993.ece Published: 27 October 2007

[5] John Vidal, *Global food crisis looms as climate change and fuel shortages bite*, The Guardian, page 15, 3/11 2007

winter harvests by nearly 40%. He warned that cereal stocks had been declining for more than a decade mainly because of climate change and now stood at around 57 days, which made global food supplies vulnerable to an international crisis or big natural disaster such as a drought or flood. He quoted Josette Sheeran, director of the UN's World Food Programme (WFP) who warned that there are 854 million hungry people in the world and 4 million more joining their ranks every year. 'We are facing the tightest food supplies in recent history.' One of the most dire consequences of climate change has to be the fact that food is being priced out of the reach of the world's most poor.

In April 2008 World Bank head, Robert Zoellick warned that 100 million people in poor countries could be pushed deeper into poverty by spiraling prices as he announced emergency measures to tackle rising food prices around the world.[6] He said inflation in the price of food was driven by increased demand, poor weather in some countries that had ruined crops and reduced production area and also because of the increase in the use of land to grow crops for transport fuels. This announcement was in response to protests and unrest in many countries including Egypt, Ivory Coast, Ethiopia, the Philippines and Indonesia over spiraling food prices. In Haiti, protests had turned violent at the beginning of the month, leading to the deaths of five people and the fall of the government.

Such is the concern with agflation that a high-level conference on world food security was convened by the UN Food and Agriculture Organization (FAO), at the beginning of June, 2008. Some 44 national leaders attended the summit in Rome, including the Japanese, French and Spanish prime ministers, the presidents of major farming nations like Brazil and Argentina and the leaders of many African nations including Zimbabwe's Robert Mugabe. On climate change, the conference Declaration states: 'It is essential to address question of how to increase the resilience of present food production systems to challenges posed by climate change... We urge governments to assign appropriate priority to the agriculture, forestry and fisheries sectors, in order to create opportunities to enable the world's smallholder farmers and fishers, including

[6] Staunton, Denis, *Food prices could lead to starvation, say IMF and World Bank*, Irish Times 14th April 2008, page 1.

indigenous people, in particular vulnerable areas, to participate in, and benefit from financial mechanisms and investment flows to support climate change adaptation, mitigation and technology development, transfer and dissemination. We support the establishment of agricultural systems and sustainable management practices that positively contribute to the mitigation of climate change and ecological balance.'[7]

On the contentious issue of biofuels, the Declaration states: 'It is essential to address the challenges and opportunities posed by biofuels, in view of the world's food security, energy and sustainable development needs. We are convinced that in-depth studies are necessary to ensure that production and use of biofuels is sustainable in accordance with the three pillars of sustainable development and take into account the need to achieve and maintain global food security...We call upon relevant inter-governmental organizations, including FAO, within their mandates and areas of expertise, with the involvement of national governments, partnerships, the private sector, and civil society, to foster a coherent, effective and results-oriented international dialogue on biofuels in the context of food security and sustainable development needs.'[8] Safeguarding food security and controlling agflation will be a major challenge in the years ahead. A failure to meet this challenge will have dire consequences.

The Death of Life

Mc Donagh refers to Professor Lonnie Thompson of Ohio State University, a world expert on glaciers who says that 4 billion people, two-thirds of the world's population, depend on water from tropical glaciers.[9] 70% of the water of the Ganges, for example, comes from melted ice. Because of floods and droughts, agricultural production in a continent like Africa, where crops are rain fed rather than irrigated, could be down 50%, this at a time of

---

[7] *World Food Summit, Rome Declaration on World Food Security,* *http://www.fao.org/docrep/003/w3613e/w3613e00.HTM , retrieved 6th June 2008*

[8] *Ibid.*

[9] Sean Mc Donagh, *Climate change, one of the most serious moral issues of our times*, J/P Alert, July/August 2007, http://www.cmsm.org/CMSM_Alert/JulAug07, Retrieved 10/10/2007

rapid population growth. He claims that a rise of just 2 degrees Celsius could bring about the extinction of 30% of the species on the planet. Plant life will be particularly vulnerable, since plants cannot migrate to a new, more suitable ecological niche in such a short period. Marine ecosystems, especially coral reefs, which are the rainforests of the oceans, are also being destroyed by climate change.

A UN report published in September 2007 warns that wildlife and biological diversity already threatened by habitat destruction and other human-generated stresses will face new challenges from climate change.[10] It says that many ecosystems are already responding to higher temperatures by advancing towards the poles and up mountainsides. Some species will not survive the transition, and 20-30 per cent of species are likely to face an increased risk of extinction. The most vulnerable ecosystems include coral reefs, boreal forests, mountain habitats and those dependent on a Mediterranean climate. Rising sea level is a major cause of concern. The best estimate for how much further the sea level will rise due to ocean expansion and glacier melt by the end of the 21st century (compared to 1989-1999 levels) is 28-58 cm. This will worsen coastal flooding and erosion. However according to the report, larger sea-level increases of up to 1 meter by 2100 cannot be ruled out if ice sheets continue to melt as temperature rises. It says that there is now evidence that the Antarctic and Greenland ice sheets are indeed slowly losing mass and contributing to sea level rise. In addition, the oceans will also experience higher temperatures, which have implications for sea life. It will impair the ability of corals, marine snails and other species to form their shells or skeletons.

One of the direst consequences is that climate change will hit the most vulnerable. The poorest communities will be most at risk to the impacts of climate change as they have fewer resources to invest in preventing and mitigating the effects of climate change. Some of the most at-risk people include subsistence farmers, indigenous peoples and coastal populations. Africa in particular is deemed very vulnerable to climate as drought has spread and

---

[10] UNEP, The *Future in our Hands- Addressing the Leadership Challenge of Climate Change,*
http://www.unep.org/Themes/climatechange/PDF/factsheets_English.pdf
23/09/2007

intensified there since the 1970s particularly in the Sahel region. It is estimated that yields in some African countries could drop by as much as 50 per cent by 2020, and some large regions of marginal agriculture are likely to be forced out of production. Forests, grasslands and other natural ecosystems are already changing, particularly in southern Africa. It warns that by 2080 the amount of arid and semi-arid land in Africa will likely increases by 5-8 per cent.

Researchers from Britain, the US and Australia, working with teams from the UN and the World Bank warned in December 2007 that the majority of the world's coral reefs are in danger of being killed off by rising levels of greenhouse gases.[11] Their study revealed 98% of the world's reef habitats are likely to become too acidic for corals to grow by 2050. The oceans absorb around a third of the 20 billion tonnes of carbon dioxide produced each year by human activity. While the process helps to slow global warming by keeping the gas from the atmosphere, in sea water it dissolves to form carbonic acid which leads to carbonation and the killing of the coral.Dr Richard Aronson who compiled a study on coral reefs and said that 'in his own life-time studying underwater landscapes (reefs) now look like hell.'[12].

Meanwhile, a fifth of Ireland's native plant life is under threat because of global warming. These figures come from a study published by the National Botanic Gardens in November 2007. According to the research 171 species of flora could be facing extinction by 2050.[13].

Three quarters of all of Europe's nesting bird species are likely to suffer declines in range due to the impact of human-induced climate change, according to Climatic Atlas of European Breeding Birds, jointly written by scientists from Durham University, the University of Cambridge and the Royal Society Bird Protection

---

[11] Ian Sample, *Acidic seas may kill 98% of world's reefs by 2050*, The Guardian, 14 December, 2007, page 14.
http://www.guardian.co.uk/environment/2007/dec/14/carbonemissions.climatechange

[12] Sean Mc Donagh, *The Death of Life-The Horror of Extinction*, Columba Press, 2004, Page 43

[13] *Aine Kerr, A fifth of plant life faces climate change risk, page 13, Irish Examiner 30/11/07*

(RSPB).[14] The estimates used in the atlas are based upon a model of climatic change which projects an increase of global average temperature of about three degrees centigrade since pre-industrial times. Oran O 'Sullivan, chief executive of Bird Watch Ireland, said the Atlas indicated that Ireland may stand to lose some iconic species such as the corncrake forever. He said that the 'problem for these birds is that they will not be adaptable to the pace of change' brought about by global warming.[15]
The ecological impact of climate change is such that the web of life upon which humans depend for survival is slowly but surely being extinguished 'with disastrous consequences for future generations.'[16]

Climate Change and Health

Dr.Hugh Montgomery Director of University College London Institute for Human Health and Environment and member of the British Health Care Carbon Council predicts that climate and weather will have profound effects on human health in the years ahead.[17] He says that extremes of climate cycle such as the El Nino Southern Oscillation are frequently associated with disease outbreaks, just as are flooding and drought elsewhere. He expects more deaths to occur from summer heat in Western cities. Elsewhere, he believes that by 2100, the population exposed to malaria-prone temperatures may rise by more than one third.

The objective of World Health Day April 2008 organized by the World Health Organization was to catalyze public participation in the global campaign to protect health from the adverse effects of climate change. WHO aims to put public health at the centre of the UN agenda on climate change. WHO Director-General, Dr Margaret Chan used the occasion to warn that climate change is

---

[14] *Brian Huntley et al, A Climatic Atlas of European Breeding Birds, Lynx Edicions, 2008*
[15] *Seán Mac Connell, Ireland's iconic native birds at risk from climate change. The Irish Times, Wed 16 January 2008 Page 14*
[16] *Sean Mc Donagh, The Death of Life-The Horror of Extinction, Columba Press, 2004, Page 57*
[17] *Hugh Montgomery, Prognosis looks increasingly grim for the health of nations-Threats to the water supply, crop failure and mass migration are all looming, Times London, March 10, 2007, page 18*

having an impact on human health. 'The core concern is succinctly stated: climate change endangers human health,' said Dr Chan.[18] 'The warming of the planet will be gradual, but the effects of extreme weather events -- more storms, floods, droughts and heat waves -- will be abrupt and acutely felt. Both trends can affect some of the most fundamental determinants of health: air, water, food, shelter and freedom from disease.' She went on to explain that human beings are already exposed to the effects of climate-sensitive diseases and these diseases today kill millions. They include malnutrition, which causes over 3.5 million deaths per year, diarrheal diseases, which kill over 1.8 million, and malaria, which kills almost 1 million. 'Although climate change is a global phenomenon, its consequences will not be evenly distributed,' said Dr Chan. These impacts will be disproportionately greater in vulnerable populations, which include the very young, elderly, medically infirm, poor and isolated populations. Dr Chan said that the 'WHO is committed to do everything it can to ensure all is done to protect human health from climate change.'

War

At a major UN climate summit held in Nairobi in November 2006, delegates heard UN secretary general Kofi Annan criticise the 'frightening lack of leadership' in tackling global warming resulting in a threat to peace and security.[19] He explained that 'changing patterns of rainfall, for example, can heighten competition for resources, setting in motion potentially destabilising tensions and migrations... There is evidence that some of this is already occurring; more could well be in the offing.'

In April 2007, on Britain's initiative, the UN Security Council held its first debate on the effect of climate on war and conflict. The UN Secretary-General, Ban Ki-Moon, said that 'when resources are scarce - whether energy, water or arable land - our fragile ecosystems become strained, as do the coping mechanisms of groups and individuals. This can lead to a breakdown of established

---

[18] Chan, Margaret, *Climate change will erode foundations of health,* http://www.who.int/mediacentre/news/releases/2008/pr11/en/index.html, retrieved 8th April, 2008.

[19] BBC News, *UN chief issues climate warning, Wednesday,* 15 November 2006 http://news.bbc.co.uk/2/hi/science/nature/6149340.stm, retrieved 15/09/2007

codes of conduct, and even outright conflict'.[20] He warned that climate and the environment have thus become one of the threats 'to international peace and security, which the UN Security Council is meant to deal with.' He explained that a committee of prominent American military officers had recently stated that 'climate changes are a threat multiplier for instability in some of the most volatile regions of the world'.

Ole Danbolt Mjøs, Chairman of the Norwegian Nobel Committee speaking at the Nobel Prize giving ceremony in Oslo in December 2007 warned that 'global warming not only has negative consequences for 'human security', but can also fuel violence and conflict within and between states.[21] He said 'the consequences are most obvious, however, among the poorest of the poor, in Darfur and in large sectors of the Sahel belt, where we have already had the first climate war. The wind that blows the sand off the Sahara sets people and camels moving towards more fertile areas. The outcome is that nomads and peasants, Arabs and Africans, Christians and Muslims from many different tribes clash in a series of conflicts.' He noted that similar conflict over scarce resourses had spread to Chad and the Central African Republic and that parts of the Sahel belt, from the Sudan to Senegal, are coming under threat.

The Consequences for Ireland

An Environmental Protection Agency (EPA) report published in August 2007 on meteorological indicators in the Irish climate states that since 1980, temperature increases in Ireland have been double the global average. Ireland today is wetter and hotter than it was 30 years ago.[22] Regionally it means that, in climate terms, Ireland is under threat from the negative effects of global warming more than

---

[20] UN News Centre, *Statement at the Security Council debate on energy, security and climate*, 17 April 2007, http://www.un.org/apps/news/infocus/sgspeeches/search_full.asp?statID=79, retrieved 20/10/2007

[21] Norwegian Nobel Institute, *Speech given by The Chairman of the Norwegian Nobel Committee - Ole Danbolt Mjøs* (Oslo, December 10, 2007) http://nobelpeaceprize.org/eng_lect_2007a.html, retrieved 11/12/2007

[22] John Sweeney, Laura McElwain , *Key Meteorological Indicators of Climate Change in Ireland*, EPA, Aug 29 2007, http://www.epa.ie/news/pr/2007/aug/name,23318,en.html, retrieved 21/10/2007

any other European country. The report's authors, Laura McElwain and John Sweeney, work at the Irish Climate Analysis and Research Units (ICARUS) on the National University of Ireland's Maynooth campus. They collected daily temperature and rainfall data from meteorological stations across the 26 Counties over the last century and compared the patterns of Irish weather data with international indicators. International temperature trends over the last 100 years have thrown up two distinct periods of global warming. The first was from 1910 to the mid-1940s and again from 1980 to 2004. It is in the period after the 1940s that Ireland begins to present its own distinct global warming pattern. While mean temperatures globally fell after the end of the Second World War, it took Ireland longer to cool, but more important has been the second period of warming beginning in 1980. McElwain and Sweeney found that global warming in Ireland has been growing at a much faster rate than the global temperature increase. They also found that rainfall has increased on the western and northern coasts of Ireland, with more intense and longer periods of rainfall that they believe may 'provide a cause for concern as they may have a greater impact on the environment, society and economy'. For example, four of the wettest five years on record at Malin Head were recorded since 1990. The increased and more intense rainfall has implications for river flood management and infrastructure provision with the possibility of floods in the winter and water shortages in the summer. Summer droughts could lead to soil erosion. Other EPA studies have found that Ireland could face water shortages over the next 20 years, particularly in the Dublin and the south-east regions. According to the EPA, Dublin has only a single source of water and is running within 1% of capacity. While increased rainfall was one aspect of the report's findings, more heat waves was another. McElwain and Sweeney state that heat waves are 'a cause for concern' because of their impact on human health, agriculture and water supply. The number of heat waves has increased at a number of stations annually, with greater increases in winter, spring and summer heat waves. They predict that, in coming years, heat waves may increase in severity, frequency or duration.

The main conclusions of the report are that Ireland is warmer and the west south-west and north are wetter, with more frequent and more intense rainfall. The mean annual temperature has increased by 0.7° C between 1890 and 2004. Six of the ten warmest years on record have occurred since 1990. The increases in intensity and frequency of extreme precipitation events are a cause for concern. They conclude that Ireland's changing climate is going to have the greatest impact on the availability of water for domestic, industrial and agricultural use. It will also impact on farming as some crops such as potatoes will struggle to grow in drought conditions while other crops such as rapeseed will flourish. Finally we are likely to experience more extreme weather conditions in the future such as prolonged summer drought, heat wave, storms and flooding.

Effects may be underestimated

Dr Martin Manning a director of the UN's Intergovernmental Panel on Climate Change (IPCC) and one of the world's most influential climate scientists has warned that global warming is happening so fast that it may become impossible to predict its full effects in the future. He was one of the authors of the IPCC's fourth assessment report on climate change released in February 2007. Speaking at a lecture on climate change organised by the Environmental Protection Agency in Dublin in November 2007 he was quoted as saying that the inability of scientists to predict something as fundamental as the worst-case scenario for rising sea levels was 'worrying the hell' out of him.[23] He said that in 2007 alone the extent to which Arctic sea ice had melted was outside the entire computer modelling that scientists worldwide had done. He said it was even more worrying that the extent of the melting, which was 20 per cent greater than the previous record in 2005, suggested that global warming may be accelerating because of 'positive feedback'. Positive feedback occurs when the Arctic sea ice melts because of rising temperatures. Increased sunlight warms the oceans further, which in turn melts more ice, accelerating the melting of the polar ice cap and leading to both higher temperatures and rising sea levels. He warned that 'it's hard to tune a model for climate change based on the past. We are seeing phenomena with a warmer world that we have had no experience of before.'

---

[23] Ronan McGreevy, *Climate change outpacing science*, Irish Times 21.11.07, page 5

Professor Wieslaw Maslowski a researcher from the Naval Postgraduate School, Monterey, California along with researchers from NASA and the Institute of Oceanology, Polish Academy of Sciences carried out research on the melting of the Arctic sea ice. Their latest modelling studies published in December 2007 indicate northern polar waters could be ice-free in summers within just 5-6 years and that that previous projections had underestimated the processes now driving ice loss.[24] Summer melting in 2007 reduced the ice cover to 4.13 million sq km, the smallest ever extent in modern times. Remarkably, this stunning low point had not even been predicted by Professor Maslowski and his team in previous predictions.

The Intergovernmental Panel on Climate Change proposes a maximum sea level rise of 81cm (32in) this century. However the world's sea levels could rise twice as high this century as UN climate scientists have predicted, according to a study carried out by Britain's National Oceanography Centre in Southampton, the results of which were published in December 2007. The researchers found that the true maximum could be about twice that: 163cm (64in).[25] Their results concur with other studies such as those carried out at the Potsdam Institute for Climate Impact Research, Germany, showing that current sea level projections may be very conservative. Scientists in Southampton looked at what happened more than 100,000 years ago - the last time Earth was this warm at the so-called interglacial period, some 124,000 to 119,000 years ago. Their research was the first to examine potential constraint on the dynamic of the polar ice sheets melting, which the most recent IPCC Assessment Report failed to do. They predict an average sea level rise of 1.6m per century brought about by melting ice that is roughly twice as high as the maximum estimates in the IPCC Fourth Assessment Report. Stefan Rahmstorf, from the Potsdam Institute for Climate Impact Research, Germany, and colleagues plotted global mean surface temperatures against sea level rise, and found that levels could rise by 59% more than current forecasts.

---

[24] Jonathan Amos, *Arctic summers ice-free by 2013*, http://news.bbc.co.uk/2/hi/science/nature/7139797.stm, Retrieved 12 December 2007

[25] BBC, *Rising seas to beat predictions*, http://news.bbc.co.uk/2/hi/science/nature/7148137.stm, Retrieved 18/12/2007

In 2005 Sergei Kirpotin at Tomsk State University in western Siberia and Judith Marquand at Oxford University discovered that an area of permafrost spanning a million square kilometres - the size of France and Germany combined - had started to melt for the first time since it formed 11,000 years ago at the end of the last ice age. The area, which covers the entire sub-Arctic region of western Siberia, is the world's largest frozen peat bog and scientists fear that as it thaws, it will release billions of tonnes of methane, a greenhouse gas 20 times more potent than carbon dioxide, into the atmosphere. It is a scenario climate scientists have feared since first identifying 'tipping points' - delicate thresholds where a slight rise in the Earth's temperature can cause a dramatic change in the environment that itself triggers a far greater increase in global temperatures. Dr Kirpotin warned the situation was an 'ecological landslide that is probably irreversible and is undoubtedly connected to climatic warming'.[26]

The fact that scientists may be underestimating the effects of climate change is now becoming a major cause for concern.

Conclusion

Al Gore, speaking at the ceremony presenting him with The Nobel Peace Prize in December 2007 more or less summed up the consequences of climate change. He said:

> In the last few months, it has been harder and harder to misinterpret the signs that our world is spinning out of kilter. Major cities in North and South America, Asia and Australia are nearly out of water due to massive droughts and melting glaciers. Desperate farmers are losing their livelihoods. Peoples in the frozen Arctic and on low-lying Pacific islands are planning evacuations of places they have long called home. Unprecedented wildfires have forced a half million people from their homes in one country and caused a national emergency that almost brought down the government in another. Climate refugees have migrated into areas already inhabited by people with different cultures, religions, and traditions, increasing the potential for conflict. Stronger storms in the Pacific and Atlantic

[26] Ian Sample, *Warming hits 'tipping point'-Climate change alarm as Siberian permafrost melts for first time since ice age*, The Guardian, August 19 2005, page 19

> have threatened whole cities. Millions have been displaced by massive flooding in South Asia, Mexico, and 18 countries in Africa. As temperature extremes have increased, tens of thousands have lost their lives. We are recklessly burning and clearing our forests and driving more and more species into extinction. The very web of life on which we depend is being ripped and frayed. [27]

It took until 1830 for the world's population to reach one billion and a further one hundred years for it to double to two billion. Eighty years later approaching seven billion of us share the planet. In the past low-density populations were able to adapt to changes in climate by moving to where the food and water were. Today, more than three billion live in cities, depending upon a complex infrastructure to provide power, food, and water.The web of life upon which we depend is slowly but surely being made extinct. While climate change will have an impact on Ireland, with proper planning these effects can be overcome. In fact the advantages might outweigh the disadvantages as we look foward to hotter Mediterranean style summers and shorter milder winters.It is the poorest parts of the world with the lowest carbon footprints that are going to experience the most devestaing impacts. Here, changing patterns of disease, rainfall and food availability now carry more sinister threats: mass migration and war.

As I stand before a group of students in the classroom I often think about the world they will find themselves in as they reach adulthood. There is so much talk about preparing young people for the future world of work and none about the manner in which global warming is likely to impinge upon their livlihoods and welfare. Those of us who are adults are unlikely to experience the worst outcomes of global warming. Nevertheless there is a huge onus on everone in the field of education to educate our young people about the likely consequences of climate change and prepare them for the likely outcomes.. This can only happen if there is a committment made to the ongoing training and upskilling of teachers in the whole area of education for sustainable

[27] The Norwegian Nobel Institute, 'The Nobel Lecture given by The Nobel Peace Prize Laureate 2007', Al Gore (Oslo, December 10, 2007) http://nobelpeaceprize.org/eng_lect_2007c.html

developement. The aim of our education system should be to prevent further degradation of our planet and teach young people to contribute to a sustainable society where the value of caring for our planet is upheld.

'Climate change: I say the debate is over. We know the science, we see the threat and we know that the time for action is now.'

Arnold Schwarzenegger

Chapter 3

Introduction

Television naturalist Sir David Attenborough, speaking to the House of Commons environment committee in Westminster in December 2006 said that global warming would worsen. 'What we can do is make the situation deteriorate less than it's going to.'[1] He went on to tell them that if no action was taken on emissions, there was more than a 75% chance of global temperatures rising between two and three degrees Celsius over the next 50 years. Sir David confessed to MPs that he had had doubts about climate change until attending a lecture by a US expert which proved that recent climate change was man-made, rather than part of the cycle of nature. Now, he added, he was sure there was 'not only climate change but humanity is responsible for that.' Sir David's address to the House of Commons reveals the extent to which politicians are beginning to take the issue of Climate Change seriously.

Scientists have coined a term for our new age - they call it the 'anthropocene' because human interference with planetary systems is affecting the very life-support systems we depend upon.[2] They warn that we may be the last generation to live in an age of climate stability and that we are now entering an era outside human experience. The latest report by the Intergovernmental Panel on Climate Change (IPCC) warns that anthropogenic warming could lead to some impacts that are abrupt or irreversible, depending upon the rate and magnitude of the climate change. This Chapter reviews the political response to climate change from the Kyoto Protocol in 1997 to the Bali Roadmap in 2007 and examines the likely impact this will have on Schools.

---

[1] 'Attenborough urges moral change,' http://news.bbc.co.uk/2/hi/uk_news/politics/6176847.stm 13/12/06

[2] Steve Connor, *Climate change causes new 'epoch'*, London Independent, Saturday, 26 January 2008, http://www.independent.co.uk/environment/climate-change/climate-change-causes-new-epoch-774293.html

Kyoto Protocol

The Kyoto Protocol is a United Nations treaty made under the United Nations Framework Convention on Climate Change (UNFCCC) that was introduced in December 1997 and came into force in February 2005 with the aim of achieving the stabilization of greenhouse gas concentrations in the atmosphere at a level that would prevent dangerous anthropogenic interference with the climate system. (Article 3.7). [3] 175 countries ratified the Protocol. Of these, 36 countries and the EEC (classified as Annex I) are required to reduce greenhouse gas emissions below levels specified for each of them in the treaty. The other countries who ratified the protocol have no obligation beyond monitoring and reporting emissions. The United States although a signatory to the Kyoto Protocol, has not ratified it. Therefore it is non-binding until they ratify it and for the moment they are the single largest emitters of carbon dioxide. Australia didn't ratify it until December 2007 when Kevin Rudd the newly elected Prime Minister signed up to it on his first day in office.

Countries that fail to meet their Kyoto obligation will be penalized. Annex I countries have to reduce their greenhouse gas emissions by a collective average of 5% below their 1990 levels by 2008-2012. For many countries, such as the EU member states, this corresponds to some 15% below their expected greenhouse gas emissions in 2008. While the average emissions reduction is 5%, national limitations range from 8% reductions for the European Union as a whole to a 13% emissions increase for Ireland above its 1990 levels. The Protocol includes 'flexible mechanisms' which allow *Annex I* economies to meet their greenhouse gas emission limitation by purchasing carbon credits from elsewhere. In 2007, the carbon trading market was worth E45 billion globally and this is expected to rise to E400 billion by 2020.[4]

This facility has led to widespread criticism among those concerned with the ethics of Climate Change. Northcott for example states that 'Carbon trading is leading to a new form of imperialism,

---

[3] UNFCCC, 'Kyoto Protocol', http://unfccc.int/kyoto_protocol/items/2830.php

[4] Owen Gaffney, *Carbon trading: A new global commodity*, Irish Examiner 7/01/08,page 13

"carbon colonialism", which effectively commodifies the earth's atmosphere and forests, privatising a common resource in the monetised form of carbon credits to be traded between governments and corporations..... The current scheme simply rewards the heaviest polluters with a large injection of newly-created public wealth in permits to pollute atmospheric carbon sinks, even although they have already profited for decades from this same common good without having paid for it.'[5] He suggests an alternative to either personal or corporate emissions trading - 'the adoption by governments of a *Pigovian* approach, with the introduction of taxation on greenhouse gas production. Currently industrial economies tax productive activities by corporations and workers. Thus corporation taxes tax profits, while income and payroll taxes tax employment and human work. Traditional pre-modern economies by contrast taxed not work, which can be productive without necessarily burdening the planet ecologically, but physical commodities.'[6] He goes on to explain that a tax on the land and its product was the first historical form of taxation, as recorded in the Joseph saga of the Old Testament. 'A shift of the tax burden from employment and profits to carbon, by means of a carbon tax, would then have significant traditional precedent. Such a shift would also be the most effective means for shrinking the carbon footprint of advanced industrial economies in the short time span of twenty years, which is all the time many scientists now believe humanity has to mitigate climate change before the earth proceeds towards a runaway climate disaster.'[7]

Northcott acknowledges that 'the Kyoto Protocol represents the most significant international treaty ever contracted between peoples across the globe.' However, he feels that the present targets and policies emanating from the protocol are grossly inadequate. 'The peoples of Africa and South Asia are already bearing an economic and physical burden of adapting to severe climate change which threatens their very survival, while the North continues to promote schemes for trading pollution permits as a way of avoiding real reductions in carbon emissions. The gap between the air-conditioned conference rooms where the parties negotiate and the

---

[5] Michael Northcott, A Moral Climate: The Ethics of Global Warming, Darton,Longman & Todd Ltd, 2007, page 139
[6] Ibid, page 141
[7] Ibid, page 141

reality of a warming world symbolises the conflict between the vision of an expanding international economic order governed exclusively by market considerations and the limited carbon sinks of the earth system.'[8]

Stern Report

The Stern Review on The Economics of Climate Change, a 700-page report released in October 2006 by former World Banker and economist Nicholas Stern for the British government discusses the effect of climate change and global warming on the world economy.[9] Even though the effects of climate change had been well aired post Kyoto, the Stern Report marked the first time a Western Government weighed in with evidence that if climate change wasn't tackled the process would become irreversible.

The Report claims that there is still time to avoid the worst impacts of climate change if strong action is taken now. It claims that the scientific evidence is now overwhelming that climate change is a serious global threat, and it demands an urgent global response. It urges that the benefits of strong and early action far outweigh the economic costs of not acting because climate change will affect the basic elements of life for people around the world. Hundreds of millions of people could suffer hunger, water shortages and coastal flooding as the world warms. Using the results from formal economic models, the Review estimates that if no action is taken, the overall costs and risks of climate change will be equivalent to losing at least 5% of global GDP each year. If a wider range of risks and impacts is taken into account, the estimates of damage could rise to 20% of GDP or more. In contrast, it claims that the costs of reducing greenhouse gas emissions to avoid the worst impacts of climate change can be limited to around 1% of global GDP each year.
Stern warns that without action now and over the coming decades there is a risk of major disruption to economic and social activity, on a scale similar to those associated with the great wars and the economic depression of the first half of the 20th century. He argues

---

[8] Ibid, page 159
[9] Nicholas Stern, *Review on the economics of climate change*, October 2006, http://www.hmtreasury.gov.uk/independent_reviews/stern_review_economics_climate_change/stern_review_report.cfm

that climate change is a global problem; therefore the response to it must be international. He argues that even at more moderate levels of warming, all the evidence shows that climate change will have serious impacts on world output, on human life and on the environment. All countries will be affected. The most vulnerable – the poorest countries and populations – will suffer earliest and most, even though they have contributed least to the causes of climate change. The costs of extreme weather, including floods, droughts and storms, are already rising, including for rich countries. Taking steps towards adapting to climate change is essential. He advises that it is no longer possible to prevent the climate change that will take place over the next two to three decades, but it is still possible to protect our societies and economies from its impacts to some extent. The costs of stabilising the climate are significant but manageable but delaying will be dangerous and much more costly.

The good news is that action on climate change will also create significant business opportunities, as new markets are created in low-carbon energy technologies and other low-carbon goods and services. These markets could grow to be worth hundreds of billions of dollars each year, and employment in these sectors will expand accordingly. Stern says that the world does not have to choose between averting climate change and promoting growth and development. Changes in energy technologies will create opportunities to decouple growth from greenhouse gas emissions.

The Stern Report met with both favourable and critical comment. Kenneth J. Arrow the American economist, winner of the Nobel Prize in Economics and considered one of the founders of modern neo-classical economics says critics of the Stern Review don't think serious action to limit CO2 emissions is justified, because there remains substantial uncertainty about the extent of the costs of global climate change, and because these costs will be incurred far in the future. He believes that 'Stern's fundamental conclusion is justified: we are much better off reducing CO2 emissions substantially than risking the consequences of failing to act, even if, unlike Stern, one heavily discounts uncertainty and the future.' [10]

---

[10] Kenneth J. Arrow, *The Case for Mitigating Greenhouse Gas Emissions*, Project Syndicate/ The Economists' Voice, 2007, http://www.project-syndicate.org/commentary/arrow1, retrieved 12/12/2007

On the other hand Professor Richard Tol, an environmental economist and lead author for the Intergovernmental Panel on Climate Change believes 'there is a whole range of very basic economics mistakes that somebody who claims to be a Professor of Economics simply should not make. Stern consistently picks the most pessimistic for every choice that one can make. He overestimates through cherry-picking, he double counts particularly the risks and he underestimates what development and adaptation will do to impacts.'[11]

Heiligendamm G8 Summit

Some months after the publication of the Stern Report, the G8 held a summit in Heiligendamm in June 2007. The G8 leaders noted that since their previous meeting two years earlier in Gleneagles, 'science has more clearly demonstrated that climate change is a long term challenge that has the potential to seriously damage our natural environment and the global economy.'[12] They agreed that a 'resolute and concerted international action is urgently needed in order to reduce global greenhouse gas emissions and increase energy security.' They say that tackling climate change is a shared responsibility of all, and can and must be undertaken in a way that supports growth in developing, emerging and industrialised economies. They declared that they 'are committed to take strong leadership in combating climate change.' The Heiligendamm Summit marked an important turning point in the political response to Climate Change in that it was the first time that the world's leading industrial nations acknowledged the scientific evidence being put forward by the IPCC and resolved to work together to reduce carbon emissions.

Fourth IPCC Synthesis Report

Later in the year, UN Secretary General Ban Ki-Moon launched the IPCC Synthesis Report following five days of tough political

---

[11] Simon Cox and Richard Vadon, *Running the rule over Stern's numbers*, 26 January 2007, http://news.bbc.co.uk/2/hi/science/nature/6295021.stm

[12] G8 Summit 2007, *Growth and Responsibility in the World Economy*, Summit Declaration, 7th June 2007, http://www.g-8.de/Content/DE/Artikel/G8Gipfel/Anlage/2007-06-07-gipfeldokument-wirtschaft-eng,property=publicationFile.pdf

negotiations between delegations in Valencia, Spain. Delegates agreed on the key 20-page summary for policy-makers, which outline the latest scientific knowledge on the causes and effects of climate change. This aims to provide a compass for governments, legislators and other decision-makers on how to mitigate carbon emissions and adapt to a changing climate. The report paints a stark picture of how climate change is already negatively affecting vulnerable habitats and communities. It warns of extreme consequences unless drastic action is taken soon to curb greenhouse gas emissions. Speaking ahead of the meeting, top UN official Yvo de Boer, executive secretary of the UN Framework Convention on Climate Change (UNFCCC) said that 'Climate change will hit hardest the poorest and most vulnerable countries. Its overall effect, however, will be felt by everyone and will in some cases threaten people's very survival. Failing to recognize the urgency of this message and acting on it would be nothing less than criminally irresponsible.'[13] The report was published in November 2007, two weeks ahead of the UN conference in Bali, Indonesia, which centred on creating a successor to the Kyoto Protocol and is further evidence that politicians were at last beginning to take the scientific evidence for Climate Change seriously but as yet there was no consensus on how to meet this challenge.

Bali Summit

The United Nations Climate Change Conference in Bali in December 2007, hosted by the Government of Indonesia, took place at the Bali International Convention Centre and brought together more than 10,000 participants, including representatives of over 180 countries together with observers from intergovernmental and nongovernmental organizations and the media. The purpose of the summit was to reach agreement on a replacement for the Kyoto Protocol which expires in 2012. The two-week period included the sessions of the Conference of the Parties to the UNFCCC, its subsidiary bodies as well as the Meeting of the Parties to the Kyoto Protocol. A ministerial segment in the second week concluded the Conference. The conference culminated in the adoption of the *Bali Action Plan*, which charts the course for a new negotiating process

[13]Marlowe Hood, *Failure to tackle global warming "criminal"*, Irish Examiner, Tuesday Nov 11th 2007, page 5

to be concluded by 2009 that will ultimately lead to a post-2012 international agreement on climate change.[14]
The Bali Action Plan accepts the findings of the Fourth Assessment Report of the Intergovernmental Panel on Climate Change that warming of the climate system is unequivocal, and that delay in reducing emissions significantly constrains opportunities to achieve lower stabilization levels and increases the risk of more severe climate change impacts. It accepts that deep cuts in global emissions will be required to achieve the objective of stabilising the climate. It agrees to long-term co-operative action to reduce greenhouse gas emission including the launch of an Adaptation Fund.

The Bali Action Plan got a mixed reception. Keith Allot, Head of Climate Change, World Wildlife Fund-UK commenting on the outcome of the talks in Bali stated that 'it's not the end of the line and there will still be a lot of work needed in order to get firm caps put in place at Copenhagen in 2009, but everyone who has ratified the Kyoto deal are clearly showing that they recognise what is required of them, that they are listening to the science and that they understand what they need to do in the years after Kyoto.'[15] He went on to state that 'it's an extremely positive indication of what the world is capable of doing when the Bush administration is removed from the table. It's an inspiration for leading the roadmap forward and shows that if we all stand together - the progressive rich nations, shoulder to shoulder with the developing countries and the emerging giants of China and India - that we can make the important decisions that need to be made.'
The outcome of the summit was met with a more negative reaction from Greenpeace. 'The Bush Administration has unscrupulously taken a monkey wrench to the level of action on climate change that the science demands,' said Gerd Leipold, Executive Director of Greenpeace International. 'They've relegated the science to a footnote.' [16] The Bali Summit marked a major turning point in the political response to climate change. There was at last a global

---

[14] UNFCC, *Bali Action Plan*, 14th December 2007, http://unfccc.int/meetings/cop_13/items/4049.php
[15] Keith Allot, *WWF Response to Bali,* 15/12/2007, http://www.wwf.org.uk/climatechange/bali_0000004594.asp#17d
[16] Ian Sample, *Deal agreed in Bali climate talks*, December 15 2007, http://www.guardian.co.uk/environment/2007/dec/15/bali.climatechange

consensus, at a political level, to the reality of climate change. The USA was at last brought in from the cold after days of intense negotiations. While many would argue that the outcome was too little, too late, it was at least a step in the right direction.

Political response to Climate Change in Ireland

The Government launched the *National Climate Change Strategy* in November 2000 and amended it in April 2007 in an effort to meet Ireland's obligations under the Kyoto Protocol. The Taoiseach Bertie Aherne acknowledged that the debate about climate change was over and stated that it was 'among the greatest challenges of our time.'[17] The *Executive Summary* states 'there is now a scientific consensus that global warming is happening, that it is directly related to man-made greenhouse gas emissions, and that we have little time remaining to stabilise and reduce these emissions if we are to avoid devastating impacts on our planet.'[18] It goes on to warn that there is 'an economic consensus that the costs of inaction will greatly outweigh the costs of action, and that progressive climate change policies, based on innovation and investment in low-carbon technology, are consistent with global economic growth..

The government intends meeting its commitments under Kyoto by reducing current Greenhouse gas emissions by over 17 million tonnes per annum, approximately 20% of which will be from the purchase of credits.[19]

The Government has also 'decided that it will use the Kyoto Protocol flexible mechanisms to purchase up to 3.607 million Kyoto Units in respect of each year of the 2008-2012 periods. This requirement will be revised as necessary in light of future projections and the impact of any additional measures to reduce greenhouse gas emissions.'[20] The Government has designated €270 million for investment in the flexible mechanisms under the

---

[17] Bertie Aherne, 'Ireland National Climate Change Strategy 2007-2012', *Foreword,* http://www.environ.ie/en/Environment/Atmosphere/ClimateChange/NationalClimateChangeStrategy/PublicationsDocuments/FileDownLoad,1861,en.pdf
[18] Ibid, page 7
[19] Ibid, page 17
[20] Ibid, page 56

National Development Plan 2007-2013. This is in addition to an initial investment of €20m in 2006. [21] San-jeev Kumar climate change expert with the World Wildlife Fund, Ireland accuses the Irish government of 'investing in hot air' rather than cleaning up its environment and helping in the fight against climate change. Ireland's purchase of carbon credits from eastern European countries, and Russia, although legal, will not reduce CO2 emissions.[22]

Following the general election in June 2007 the Green Party and Fianna Fail formed a coalition government. Mr. John Gormley, T.D., the leader of the Green Party was appointed Minister for the Environment, Heritage & Local Government. On climate change and energy in the agreed Programme for Government, the Green Party successfully negotiated for a carbon tax, an annual greenhouse gas emission reduction target of 3 percent and the establishment of national green building standards.[23]

In November 2007 the Government launched a €15 million euro campaign that aims to raise awareness about climate change and help businesses to substantially reduce their carbon emissions over the next five years.[24] It's part of an attempt to help reduce the Government's €290m bill for carbon credits and help it meet Kyoto targets. At the same time a new Cabinet committee on climate change and energy security was established to oversee the implementation of the national climate change strategy Minister Gormley said the 'the central theme of this Change Plan of Action is a challenge to the Irish Nation to change the way we think about climate change when it comes to travel, work, business, home and leisure. We all have a vested interest in succeeding in this objective and we all have a responsibility to play our part. Doing nothing is not an option. Our only option is to change how we think about

---

[21] Ibid, page 57

[22] Ann Cahill, *Hot air; report criticises Irish plan to buy carbon credits*, Irish Examiner 13th June 2007, page 13

[23] RTE, *Green party members voting on deal*, http://www.rte.ie/news/2007/0613/election.html , retrieved 13th June 2007

[24] Department of the Environment, Heritage and Local Government, Press Release, *Government makes urgent call of action to Irish Society to tackle climate change*, 29/11/2007, http://www.environ.ie/en/Environment/Atmosphere/ClimateChange/News/MainBody,16169,en.htm

climate change and change our behaviour to quickly embrace the reality that we are living in a world in which emissions of greenhouse gases are being progressively restricted.'[25] A few weeks later he delivered the first Carbon budget in the history of the State. It aims to reduce carbon emissions by an average of three per cent per anum over the next five years. It's hoped measures announced in the budget including green energy grants, insulation grants and VRT reforms will lead to further reductions of an average of 600,000 tonnes per anum between 2008 and 2012.[26]

Figures released by the Environmental Protection Agency (EPA) in January 2008 show that Ireland's total greenhouse gas emissions in 2006 of 69.77 million tonnes were 25.5% above 1990 levels.[27] Under the Kyoto Protocol Ireland is allowed a 13% increase on 1990 levels, which amounts to 62.84 million tonnes. This means Ireland's emissions in 2006 were almost seven million tonnes above the limit. Commenting on the figures Dr Mary Kelly, Director General, EPA said that 'while the figures are encouraging, and the reduction of 0.8% on 2005 levels is most welcome, the remaining distance to our Kyoto target is substantial and shows that we continue to face a very major challenge. Reducing emissions in a growing economy will require a major effort on all our parts.'[28] She went on to explain that the rise of transport emissions was by far the largest in any sector in 2006 and reflected a 165 per cent increase on 1990 figures. Transport emissions made up almost 20 per cent of the 2006 total, most of which were generated by road transport (97%). Commenting on whether Ireland can reach its Kyoto target by 2012, Dr Kelly said 'that the Government's target of 3% annual reductions in emissions over the next 5 years will be extremely challenging and further emphasises that actions to reduce domestic emissions must be intensified and strengthened. Greenhouse gases emitted now will remain in the atmosphere for

---

[25] Ibid

[26] Department of the Environment, Heritage and Local Government, Press Release, *Minister Gormley delivers State's first carbon budget* http://www.environ.ie/en/Environment/News/MainBody,16215,en.htm 8.12.07

[27] EPA, *Ireland's Emissions of Greenhouse Gases for the period 1990-2006*, 15/01/2008
http://www.epa.ie/downloads/pubs/air/airemissions/name,23960,en.html

[28] EPA Press Release, *Overall decrease in Ireland's Greenhouse Gases in 2006*, Date released: Jan 15 2008 http://www.epa.ie/news/pr/2008/name,23984,en.html

many decades and affect the climate for centuries to come.' In reality the Irish government's approach to tackling climate change is to avail of the Emissions Trading Scheme. The current budget designates €270 million to purchase carbon credits. This attitude fails to address the core problem of reducing carbon emissions.

In April 2008 Minister Gormley, launched the consumer communication aspect of the Government's climate change campaign. The '*Change*' campaign has two key and overriding imperatives; firstly to change how people in Ireland think about climate change, and secondly to encourage everyone to change how they behave. The communications campaign includes a comprehensive website, www.change.ie, a lo-call information line and advertising, all of which are backed up by an extensive stakeholder engagement process that is working with all of the sectors to change behaviour and reduce Ireland's greenhouse gas emissions with a view to halting climate change. Market research conducted by Behaviour & Attitudes on behalf of the 'Change' campaign has found that 82% of people are concerned about the impact of climate change on Ireland. The Change campaign encourages everyone to change how they behave, at home, at work and in the community. Whether or not this will have any impact on government efforts to meet our commitment under the Kyoto Protocol remains to be seen.

Conclusion

The political response to climate change from Kyoto to Bali is a fifteen yearlong saga which resulted in a considerable change in the level of consciousness among politicians in relation to climate change. The most significant transformation is that the debate among politicians over whether or not climate change is happening is over. There is consensus now among all political leaders that it's taking place and that human activity is the principle cause. The signing of the Fourth IPCC Synthesis Report in Valencia in Spain ahead of the Bali Summit was a key moment in this turnaround. In time, The Bali Summit will probably be seen as a turning point in history when the political representatives of almost every nation on earth accepted the evidence being put forward by the scientific community regarding climate change. The scientific evidence for climate change has to be incredibly compelling if the USA, the greatest sceptic of them all has been convinced. The journey from Kyoto to Bali is symbolic of a conversion among political leaders from unbelief to belief in the scientific arguments for climate change. At home, the government appears to be making more of an effort to confront the issue. However it is worth noting that An Taoiseach, Brian Cowen has been more or less silent on the issue since he came to office in May 2008. Shortly after becoming Taoiseach he addressed the National Forum on Europe, held in Dublin Castle, but made no reference to climate change.[29] He did however manage to cover a wide range of other issues from sovereignty to foreign investment.

In June 2008, Prime Minister Gordon Brown backed a call for more global learning to take place in schools across England at the launch of *Global Matters* a new Development Education Alliance (DEA) publication which was launched in Westminster by DFID Minister Shahid Malik MP, and QCA Director of Curriculum Mick Waters. The Prime Minister in the foreword emphasises his commitment to young people learning about global issues and the impact that their own actions has on people. He says, 'I want to see the teaching of global issues given more weight in our schools and colleges - and already we have taken steps to make issues like globalisation, environmental sustainability and citizenship a core part of the curriculum. For it is only through education that we will foster citizens with the conviction to speak out against world poverty, that we will find the creativity we need to tackle climate change and that we will produce the next generation of social entrepreneurs.'[30] The Department for Children, Schools and Families is one of three new government departments set up by the Prime Minister in June 2007 and education for sustainable development is part of its remit. It is developing policies and programmes for children, schools and families that promote and support sustainability in the classroom. This is part of a government effort to make all schools in Britain sustainable by 2020. Elsewhere, *Building Schools for the Future* is an immensely ambitious programme designed to rebuild or refurbish all secondary schools in England over 15 years at a cost of £45 billion. Education for sustainable development is on a statutory footing and there is a commitment to fund it. Climate

---

[29] Cowen, Brian, *Address to the National Forum on Europe*, Dublin Castle, Thursday, 22nd May 2008, http://www.taoiseach.gov.ie/index.asp?locID=582&docID=3896

[30] DEA, Prime Minister backs DEA call for more funding, 12th June 2008, http://www.dea.org.uk/news-5d7f7af56302e8ef7063f869e94d2ca7

change is a key element in the programme. To date, nobody in the Irish government has the foresight to see that one of the places to start tackling the issue of climate change is in our schools. It would appear that no effort is being made by our government to make a similar commitment to education for sustainable development in this country. If the Irish government is really serious about tackling climate change and meeting our commitments under the Kyoto Protocol, then it will have to co-ordinate the various government departments in a drive to promote education for sustainable development in our schools. David Attenborough appears to have stirred the House of Commons into action, perhaps a similar speaker addressing both Houses of the Oireachtas might rouse our government representatives into putting education for sustainable development on a statutory footing.

'The danger posed by war to all of humanity - and to our planet - is at least matched by the climate crisis and global warming. I believe that the world has reached a critical stage in its efforts to exercise responsible environmental stewardship.'

UN Secretary General Ban Ki-moon

Chapter 4

Introduction

The Catholic Church has had an ambivalent attitude towards earth and creation, and theologian Dr McDonagh is critical of church leaders who remain silent on the massive problems that face the earth. He points out that the first reference to global warming in papal teaching is in the 1990 document on ecology entitled *Peace with God the Creator, Peace with All Creation.*[1] The next mention of climate change comes in 2004 in the *Compendium of the Social Doctrine of the Church* and merits only a single article, No. 470 which according to him 'fails to capture either the magnitude of the problem or the urgency with which it must be faced.'[2]

In November 2007 Pope Benedict published his New Year's Day message for 2008 to coincide with the UN Summit on climate change held in Bali. He states that 'human beings, obviously, are of supreme worth vis-à-vis creation as a whole. Respecting the environment does not mean considering material or animal nature more important than man. Rather, it means not selfishly considering nature to be at the complete disposal of our own interests, for future generations also have the right to reap its benefits and to exhibit towards nature the same responsible freedom that we claim for ourselves.'[3] The Pope's message goes on to state that 'humanity today is rightly concerned about the ecological balance of tomorrow. It is important for assessments in this regard to be carried out prudently, in dialogue with experts and people of wisdom, uninhibited by ideological pressure to draw hasty conclusions, and above all with the aim of reaching agreement on a model of sustainable development capable of

---

[1] Sean McDonagh, 'Climate change, one of the most serious moral issues of our times', Conference of Major Superiors of Men Justice and Peace office.,*J/P Alert*, July/ August 2007, http://www.cmsm.org/CMSM_Alert/JulAug07/ retrieved 10/10/2007

[2] Ibid.

[3] His Holiness Pope Benedict XVI, The Human Family a Community of Peace, Message for the Celebration of World Day of Peace, 1 January 2008, paragraph 7 http://www.vatican.va/holy_father/benedict_xvi/messages/peace/documents/hf_ben-xvi_mes_20071208_xli-world-day-peace_en.html, retrieved 15/12/2007

ensuring the well-being of all while respecting environmental balances.'[4] Nowhere in the message is the word *climate change* mentioned. This led to ambiguity and widespread misinterpretation of what he said. The following day's edition of the *Daily Mail* ran a headline: *The Pope condemns the climate change prophets of doom.* The article stated that: 'The leader of more than a billion Roman Catholics suggested that fears over man-made emissions melting the ice caps and causing a wave of unprecedented disasters were nothing more than scare-mongering.'[5] It beggars belief that the Head of the Catholic Church didn't make a specific statement on climate change at a time when the representatives of over 180 countries were gathered together to work towards a universal solution to a problem of life threatening magnitude. Perhaps the pope's failure to respond adequately to the crisis can be explained by a dualistic approach to *human beings* and *creation.* Sean Mc Donagh believes that this is one of the main causes for the current ecological crisis because the church in the past has tended to make a distinction between matter and spirit.[6] Consequently theologians have been concerned almost exclusively with the process of salvation and sacred matters and have for the most part ignored threats to the natural world.

The Department of Education and Science in its *Guidelines for Teachers of Leaving Cert Religious Education* points out that the Catholic Church has been 'slower to embrace environmental concerns' than the World Council of Churches (WCC) who have a 'concern for the integrity of creation' as part of their agenda 'for the past three decades.'[7] It notes that to date, the Irish Catholic Bishops have not issued a pastoral letter on the environment. At the June 2007 General Meeting of the Irish Bishops it was announced that a Bishops' Conference pastoral letter on the environment would take place in the coming months which would specifically

---

[4] Ibid, paragraph 7

[5] Simon Caldwell, *The Pope condemns the climate change prophets of doom*, Daily Mail, 14.11.07, http://www.dailymail.co.uk/pages/live/articles/news/worldnews.html?in_article_id=501316&in_page_id=1811 retrieved 15/12/2007

[6] Sean Mc Donagh, *The Death of Life-The Horror of Extinction*, Columba Press, 2004, pages 58-59

[7] Department of Education and Science, *Guidelines for Teachers of Leaving Cert Religious Education*, 2004, Page 63

address climate change.[8] This has not happened. Instead, the Trocaire 2008 Lenten Campaign was used to highlight the issue. Meanwhile the bishops chose alcohol abuse as their Lenten theme. Trócaire is the official overseas development agency of the Catholic Church in Ireland and it provides excellent recourses for schools on the issue of global warming.[9]

This chapter examines the ethics of climate change and how the World Council of Churches (WCC) has 'prophetically witnessed to the threat presented by global warming on behalf of those who are already suffering from global warming.'[10] The Catholic Church has shown little of no leadership in this area and this has implications for Irish schools, the vast majority of which are under its patronage.

More than 3000 of the 3,200 primary schools in the Republic of Ireland are controlled by the Catholic Church. Over half of the 732 second level schools are religious run and among these are the CEIST schools. CEIST (Catholic Education, an Irish Schools Trust) is a new trustee body for the voluntary secondary schools of the Daughters of Charity, the Presentation Sisters, the Sisters of the Christian Retreat, the Sisters of Mercy and the Missionaries of the Sacred Heart and has trusteeship of 112 second level schools with 58,000 students and 4,000 teachers and administrative staff. One in six of all those attending second level in the Irish Republic and almost a third of those in faith-based secondary schools are now under the control of this new body. Although climate change is not specifically mentioned in the CEIST *Charter*, it is implicit in the statement that 'the call of Christian discipleship demands opening our eyes to the reality of life, feeding those who are too weak to feed themselves, liberating those who are oppressed, expanding our minds through education unsealing our ears to hear the divine echo in our hearts and inspiring hope for the future.'[11] It says that

---

[8] Catholic Communications Office, Press Release, 'June General Meeting of the Irish Bishops Conference in Maynooth,' 13th June 2007, http://www.catholiccommunications.ie/Pressrel/13-june-2007.htmlBishops , retrieved 1/08/2007

[9] Trocaire, *School Resources*, https://www.trocaire.org/lent/schools/resources.php, retrieved 11/06/2008

[10] Michael Northcott, *A Moral Climate, the ethics of global warming*, page 42

[11] CEIST, 'Charter', page 6, http://www.ceist.ie/index.php/plain/charter/charter, 12/12/2007

'CEIST engages with all people of good will to promote a preferential option for those made poor, to take action for justice, and to exercise care of the earth in a spirit of respect and welcome for diversity.'[12] Ultimately, 'A CEIST school actively encourages all its members – staff and students – to reflect on the contemporary world in the light of the Gospel.'[13] I will show that there is a problem for Catholic schools like those run by CEIST because the Catholic Church has not yet published a pastoral letter on climate change nor have they issued a statement stressing the urgency of the matter for those concerned with education. This is a serious matter as aforementioned, most Irish children attend a Catholic school.

Prophetic Leaders- The World Council of Churches (WCC)

The World Council of Churches (WCC) represents 560 million Christians worldwide. It established a working group on climate change in the run up to the UN Rio Earth Summit in 1992 which has been the facilitator of its climate change programme ever since. Sean Mc Donagh says that the WCC has given the most courageous leadership of any Christian institution on the issue of climate change.[14] The WCC made a statement to the High-Level Ministerial Segment of the UN Climate Conference in Nairobi in November 2006.[15] It said that 'caring for life on Earth is a spiritual commitment' and that 'people and other species have the right to life unthreatened by human greed and destructiveness.' However, 'human-induced climate change' caused by 'energy-intensive wealthy industrialised countries' is leading to major climate change. It is 'poor and vulnerable communities in the world and future generations will suffer the most.' It went on to warn that rich nations bear the primary responsibility for causing climate change and that they must therefore adopt strategies to drastically reduce their emissions and pay that ecological debt to other peoples by

---

[12] Ibid, page 11

[13] Ibid, page 13

[14] Sean Mc Donagh, *Climate Change, the Challenge to all of us*. Page 155

[15] Dr. Jesse Mugambi, *Climate justice for all*, A statement from the World Council of Churches (WCC) to the High-Level Ministerial Segment of the UN Climate Conference in Nairobi COP12/MOP2) November 17, 2006 http://www.oikoumene.org/en/resources/documents/wcc-programmes/justice-diakonia-and-responsibility-for-creation/climate-change-water/17-11-06-climate-justice-for-all.html

fully compensating them for the costs of adaptation to climate change.

At the UN Framework Convention on Climate Change held in The Hague in November 2000 the WCC said that 'the atmosphere envelops the Earth, nurturing and protecting life. In response to God's love for creation, we have a responsibility to care for the well-being of Earth and its ecological processes. Plants, animals and every member of the human family are dependent on this gift and have a right to its sustaining vitality.'[16] It says that 'the atmosphere belongs to no one. It is to be shared by everyone, today and in the future. Economic and political powers cannot be allowed to impair the health of the atmosphere nor claim possession of it.'

The atmosphere as a global commons

Thomas Aquinas described excess wealth accumulation as theft in his reflection on the eighth commandment, 'thou shalt not steal'. He suggests that if a rich man takes possession of something that was common property for his own use, and in such a way as to exclude others from using it, he steals common goods from others and therefore sins.[17] Michael Northcott describes 'the purloining of the commons of the atmosphere as a sink for the excessive emissions of the rich is a form of theft because it directly impedes the bodily sustenance of those in drought or flood-prone regions.'[18] He says that the injustice of global warming is all the more immoral, 'given that the cause of the problem is not the subsistence emissions of the poor but the luxury emissions of the wealthy.' He notes that 'that the richer individuals are, the more carbon they tend to consume. Even within Western societies there are considerable differences between the carbon emissions of the wealthy and of the relatively poor.' He says that 'luxury emissions represent moral malfeasance by the rich against the poor both nationally and internationally,

---

[16] World Council Of Churches, *The Atmosphere as Global Commons: Responsible and Caring and Equitable Sharing*, 6th Session of the Conference of the Parties (COP6) to the UN Framework Convention on Climate Change, The Hague, Netherlands, November 2000, http://www.wcc-coe.org/wcc/what/jpc/cop6-e.html, retrieved 3/10/2007

[17] Michael Northcott, *Moral Climate: The Ethics of Global Warming*, page 56

[18] Ibid, page 57

since they give rise both to local and global forms of pollution and other kinds of moral harms.' In his opinion 'beyond a certain level of emissions the wealthy should properly be held morally responsible for the damaging effects of their emissions on the poor. Once they know of the damage their emissions are doing, the rich have no excuse for not reining in their emissions, and for not compensating the poor for the damage their emissions are doing.'[19]

Sean Mc Donagh has written that 'one of the abiding tragedies and ironies in reflecting on global warming is that the poor who have contributed least to it, will suffer most.'[20] The WCC has called on countries with high per capita emissions of greenhouse gases to reduce those emissions dramatically and to work towards creating a non-carbon energy future which is sustainable and equitable. 'With one quarter of the global population, the rich industrialised countries currently generate three-quarters of the global CO2 emissions. They have the moral responsibility to substantially reduce their own emissions. The principle of polluter pays is relevant but insufficient if it means that countries can continue with their high levels of emissions if they have the resources to buy credits to meet their reduction targets. The polluter must change behaviour in order to reduce the pollution at source.'[21]

The WCC condemns emissions trading under the Kyoto Protocol as 'environmentally flawed.' The reason for this is because it is cheaper for rich countries to meet their carbon reduction targets by purchasing credits from other countries than by making emissions reductions at home. 'Thus rich industrialised countries would be able to meet their reduction targets in part through purchasing these fictitious emissions reductions. The net impact to the global atmospheric commons would not be a real reduction in emissions rendering the system environmentally flawed. Knowingly violating the criterion of ecological effectiveness would be tantamount to environmental fraud.' Northcott says that every individual human being, no matter how poor, has the right to a stable climate, and to

---

[19] Ibid, page 57
[20] Sean Mc Donagh, *Climate Change, the Challenge to all of us*. Page 171
[21] World Council Of Churches, *The Atmosphere as Global Commons: Responsible and Caring and Equitable Sharing*, 6th Session of the Conference of the Parties (COP6) to the UN Framework Convention on Climate Change, The Hague, Netherlands, November 2000

grow enough food to feed her family. 'Such a right is not a matter of human judgment and it may not be set in the balance against the benefits that others may derive from depriving some of this right; there are no aggregate sums which can remove this foundational right. It is the birthright of every human being. And the willingness of governments in the South to trade forests or lands or development projects for the excessive carbon emissions of wealthy Northern nations and corporations is perversely opposed to this birthright.'[22] He sees commodifying the atmosphere through carbon trading an example of 'structural sin.'[23] The WCC views this practice as a 'sin against God.' It declares that: 'True forgiveness is available from God but only after true repentance by the sinner. True repentance requires a conversion of the heart and a transformation of behaviour. Only then can true forgiveness be experienced. Countries with high emissions need a conversion of the heart and demonstrably new behaviour before they seek forgiveness.'[24] The first step towards doing this is to carry out an honest inventory of how our lifestyles are contributing to climate change.

Seeking The Truth

Al Gore, The Nobel Peace Prize Laureate 2007, speaking in Oslo at the award ceremony said that the threat of a climate crisis 'is real, rising, imminent, and universal.' [25]He went on to say that 'Mahatma Gandhi awakened the largest democracy on earth and forged a shared resolve with what he called *Satyagraha* - or *truth force*. In every land, the truth - once known - has the power to set us free. Truth also has the power to unite us and bridge the distance between *me* and *we*, creating the basis for common effort and shared responsibility.'

---

[22] Michael Northcott, *Moral Climate: The Ethics of Global Warming* page 160
[23] Ibid, page 153
[24] World Council Of Churches, *The Atmosphere as Global Commons: Responsible and Caring and Equitable Sharing,* 6th Session of the Conference of the Parties (COP6) to the UN Framework Convention on Climate Change, The Hague, Netherlands, November 2000

[25] The Norwegian Nobel Institute, The Nobel Lecture given by The Nobel Peace Prize Laureate 2007, Al Gore (Oslo, December 10, 2007) http://nobelpeaceprize.org/eng_lect_2007c.html

It is noteworthy that Gore places such emphasis on *truth*. When it comes to developing a climate change ethic, a number of theologians writing on the issue have placed truth at the heart of the matter. Michael Northcott defines parrhesia as 'the one who speaks the truth is also the one who is prepared to witness to the truth by living in service of truth and in solidarity with those who suffer at the hands of the powerful.' [26] He says 'this kind of truth-telling involves a preparedness to stand by words and to match deeds to words. Such truth-telling is foundational to ethics, for humans are only capable of responsible action when they acknowledge and respond to the authority of the perennial and substantial world from which all life derives, and which Christians call Creation.' He suggests 'that truthful response to global warming requires a proper accounting and confession of the intrinsic connection between global warming, modern imperialism and neoliberal global capitalism. Only when the rich confess the ecological harms, or eco-logical debts, with which they burden the poor and other species through the deregulated global economy will it be possible for the poor to gain justice and for climate debts to be justly redeemed.'[27]

Northcott draws parallels between the current crisis and the past. He sees the Bible as a key to working our way out of the present predicament we find ourselves in. 'The ancient saga of Noah is not only a powerful story of human survival in the face of a climate-related catastrophe but also a moral tale in which the flood is seen as divine punishment for a generation of humans whose wickedness was such that it affected the life of all flesh on earth, and endangered even the earth which was as a result full of violence (Genesis 6:13).'[28] He refers to 'Joseph, whose story offers a more hopeful scenario for our current climate change predicament than that of Noah: Joseph's warnings of imminent climate change were heeded by the Egyptians and prudent preparations were made which saved them, and subsequently Joseph's own family, from calamity.'[29] He warns 'that there is a particular responsibility on those who discern the links between these ancient stories and the natural signs of modern climate change to seek to challenge and

---

[26] Michael Northcott, *Moral Climate: The Ethics of Global Warming* page 40
[27] Ibid, page 43
[28] Ibid, page 72
[29] Ibid, page 75

repair the instrumentalising forms of making and consumption which have advanced the intergenerational, international and inter-species injustices of global warming.'[30]

While much has been written about the international and inter-species aspect, little has been written about the intergenerational aspect of climate change. Sean Mc Donagh points out that a climate change ethic must enshrine the rights of the generations yet to come. 'Future generations have a right to inherit a world as fertile and as beautiful as the one which we inhabit.'[31] The truth of the matter is that current western lifestyles are leading to climate change which threatens life on the planet now and for the generations to come. The tragedy is that those who have done least to cause it are the ones who suffer most. 'The church community and all humankind need to respond to the plight of the poor.' [32] We must also accept that our relationships extend beyond God and other human beings. 'It must also extend and include our relationships with all creation.' [33] This calls for a radical change in the way we live our lives.

Renouncing affluence

Sean Mc Donagh wrote that 'the church must challenge individuals and institutions which are primarily responsible for global warming to change their affluent lifestyles and their profligate use of energy.'[34] Michael Northcott goes further. He says that: 'Industrial civilisation is on the road to ruin, spiritually as well as ecologically, if its citizens continue to pursue endless distraction and luxury at the cost of their moral and spiritual health and the health of the planet.'[35] In a statement presented by the World Council of Churches to the plenary of high-level government representatives at the UN climate summit in Bali, Indonesia in November 2007 there is a call for society to move away from promoting endless growth and production of goods as well as a seemingly insatiable consumption. They say that 'it is our conviction as members of

---

[30] Ibid, page 80
[31] Sean Mc Donagh, *Climate Change, the Challenge to all of us,* page 168
[32] Ibid, page 172
[33] Ibid, page 175
[34] Ibid, page 176
[35] Michael Northcott, *Moral Climate: The Ethics of Global Warming* page 156

faith communities that a Change of Paradigm from one way of thinking to another is needed if we are to adequately respond to the challenge of climate change. It constitutes a transformation, a *metamorphosis*. This kind of movement just does not happen on its own; it must be catalysed by agents of change.'[36] They believe 'the world Faiths could be one of those catalysts. A change in paradigm appears as mandatory in the prevailing economic strategy of promoting endless growth and production of goods and a seemingly insatiable level of consumption among the high-consuming sectors of our societies. Such economic and consumption patterns are leading to the depletion of critical natural resources and to extremely dangerous implications with climate change and development. Societies must shift to a new paradigm where the operative principles are ethics, justice, equity, solidarity, human development and environmental conservation.' Their statement points out that the earth was entrusted to us but we simply cannot do whatever we want with it. 'We cannot make use of nature using it only as a commodity. We must bear in mind that our liberty does not allow us to destroy that which sustains life on our planet.'

Cultural historian Thomas Berry has devoted his career to understanding how nature seen as a commodity fails to sustain a nurturing relationship between humans and the Earth. Carbon trading is a prime example of this approach. In his major work *The Dream of the Earth*, he has traced how the growth of modern technological culture has led to a spiritual separation from the earth. For Berry, the primary problem facing humans today concerns the human attitude that we as a species are somehow essentially disengaged from the earth on which we live and that our destiny is to bend nature to our purposes. The story or myth that continues to drive this goal of human domination of the earth is a secular version of the old millennial dream of Christianity, a version in which God will rule the Earth and peace, harmony, and justice will prevail, brought about, however, through human

[36] World Council of Churches , *This far and no further: Act fast and act now*! Statement to the High-Level Ministerial Segment of the 13th Session of the Conference of the Parties – COP13 to the UNFCCC 3rd Session of the Meeting of the Parties to the Kyoto Protocol – CMP3 Nusa Dua, Bali, Indonesia Friday, December 14, 2007. http://www.oikoumene.org/en/resources/documents/wcc-programmes/justice-diakonia-and-responsibility-for-creation/climate-change-water/14-12-07-statement-to-cop13-un-climate-conference-bali.html , retrieved 22/12/2007

science and technology. But this destructive myth of a technological wonderland in which nature is bent to every human whim is turning the Earth, according to him, into a wasteland and threatening human survival. 'We should be clear about what happens when we destroy the living forms of this planet. The first consequence is that we destroy modes of Divine presence…to lose any of these is to diminish our own human presence.'[37]

Sean Mc Donagh says that 'our faith and our religious tradition have much to offer the world at this time, including the importance of simplicity, and of learning to give up some things that we want, so others may have what they need.'[38] He believes 'the right to a healthy environment requires a healthy human habitat. This precludes anything that might damage the life sustaining processes of the planet, which includes access to clean water, fresh air and fertile soils free from toxins, or hazards which would threaten human well-being. If, as we know, global warming is going to make the climatic conditions for the earth and humans extremely unpleasant, then campaigns to promote action on global warming could be fought under the banner of a right to a stable climate.'[39] The legacy of affluence is climate change. The challenge to Christians is to live in a way that has the least impact on the earth.

Cardinal Daly's book *The Minding of Planet Earth*, calls for urgent action at local, national and international level to stop the degradation of the environment. In the final chapter of the book he argues that the Bible calls on humans to act as stewards over creation. This demands that we act in accordance with the divine plan for creation and be subject to the moral law of justice and respect for the rights of others. Men and women are called to be caretakers, not masters, of the universe. We are given a 'duty of care for the planet', not 'a plunderer's licence.'[40]

Learning to tread lightly upon the planet

The laws of physics emphasises the interconnectedness and interdependence of all the elements which make up the universe of

[37] Thomas Berry, *Dream of the Earth*, Sierra Club Books, 1988, page 11
[38] Sean Mc Donagh, *Climate Change, the Challenge to all of us,* page 184
[39] Ibid, page 167
[40] Daly, Cardinal Cahal, *The Minding of Planet Earth*, Veritas, 2004

which the natural world is part. Ecological ethics is based on the fundamental principle that sees all things as relational. Humans are slowly beginning to see all of creation, the universe itself, the biosphere on earth, individual ecosystems, a living tree, a cell, or a proton as fundamentally relational and part of a network of interrelationships. This concept is expressed beautifully in John's gospel in the words of Jesus as he prays 'they may all be one.' (John 17: 20-23). Denis Edwards believes that humanity is integrally related to all of creation, all being part of the same story of the cosmos: Human beings share in this common story of creation. We have a communal heritage with all other creatures, in being born from the Big Bang, made from stardust, and brought to life within an evolutionary community. It is because of supernovas far out in space that our carbon based life-forms can exist. We are intimately interconnected with the whole life system of the planet and the complex interaction between living creatures and the atmosphere, the land and the water systems.[41] Therefore 'the human person has the particular dignity and responsibility which comes from being one in whom the universe has come to self-awareness.' [42]

Northcott says 'there are no autonomous human actions, for in a fundamental physical sense all actions are interconnected by their effects on the carbon cycle. Each individual action is an infinitesimal element in this cycle. The geochemical interconnection of all human actions and all life is a physical analogy for the Christian doctrine of the Communion of Saints.'[43] No man is an island. We are collectively responsible for what is happening to our planet and we are therefore collectively responsible for living in a way that puts it right. Having got ourselves into such a mess the job is to get ourselves out of it, or as Rasmussen says 'the task is to get from here (unsustainability) to there (sustainability).' [44] He suggests that first of all we view the earth as *Oikas*- 'a vast but single household of life' that provides a sustainable habitat for all living things. Current economic patterns centred on capitalism and globalisation must give way to

---

[41] Denis Edwards, *Jesus the Wisdom of God*, Orbis, N.Y.,1995, page 142
[42] Ibid, 143
[43] Michael Northcott, *Moral Climate: The Ethics of Global Warming* page 163
[44] Larry Rasmussen, *Earth Community Earth Ethics*, Orbis, 1996 page 88

stewardship and communal living. He demonstrates how the early Christians lived according to this principle of *oikas* living. 'Early Christians appropriated this instruction in household management. Their new communities were originally conceived as *oikoi,* themselves households of faith in this case, partly because the early church was literally a house-church movement.'[45]

Sally Mc Fague sets out to give Christians a working theology, one that can actually function in their personal, professional, and public lives that will lead to sustainable planetary living. She also uses the analogy of a house with 'house rules' to show how we might live in a sustainable way bearing in mind that the words *ecology* and *economics* have their roots in the Greek word *oikas* . The basic rule is that 'if everyone is to have a place at the table, the limits of planetary energy must be acknowledged. Energy that is consumed is not recycled, but goes into the atmosphere as carbon dioxide. The house rules of our home set limits to growth—both of our consumer desires and the size of the human population. We need, then, to become ecologically literate, to learn what we can and cannot do to our home if we are to continue to exist in a sustainable way. We must fit our little economy into the Big Economy, earth's economy, if our economy is to survive.' [46] Christians concerned about climate change are being called away from living as individuals towards communal living. This involves a paradigm shift in the way we view the earth and man's relationship with it. It is much easier to go into a state of denial and say that the scientists may be wrong or a state of despair by claiming there is no point in doing anything in Ireland given the rapid increase in emissions in the East as their economies begin to industrialise.

Despair and denial

Jim Power, economist and commentator wrote recently: 'I think we are a little bit carried away with the whole green agenda. At the end of the day, the two big carbon emitters in the world are China and the U S. The Chinese are opening new coal-powered electricity stations on a monthly basis and these polluters are sending out as much carbon in a matter of hours as we do in a full year. So all of

---

[45] Ibid, page 95

[46] Sally Mc Fague, *Life Abundant, Rethinking Theology and Ecology for a Planet in Peril,* Fortress Press, Minneapolis, 2001, page 209

the measures we could take here in Ireland will not make any real difference to total global carbon emissions. It is great to try and set a good global example, but one wonders if the behaviour of the Americans and Chinese will alter as result of what Minister Gormley is forcing us to do.'[47]

Dealing with people who are either in despair or denial about climate change is a major challenge which has been confronted by Michael Northcott, who maintains that 'the fossil-fuelled global economy is dangerous to planet earth and to human life, not just because it is warming the climate of the earth but because it is deeply destructive of the diversity and welfare of the ecosystems and human communities from which surplus value is extracted and traded across highways, oceans and jetstreams. The rituals encouraged by the recognition of global warming — turning off lights, turning down the heating, cycling or walking instead of driving, holidaying nearer to home, buying local food, shopping less and conversing more, addressing the causes of fuel poverty locally and internationally — are good because they are intrinsically right, not just because they have the consequence of reducing carbon emissions.'[48]

Sean Mc Donagh argues that the churches have a 'prophetic duty' with regard to climate change.[49] He makes the case that the Christian churches must take the lead in living in a sustainable way. Climate change 'is a global problem requiring real global solutions. But individual Catholics, parishes, Catholic schools, religious communities and church organisations can play a big part by making different choices.'[50] Thomas Cahill in *How the Irish Saved Civilisation*, writes about the central role played by Irish monks in maintaining European culture from the fifth century onwards during the Dark Ages following the collapse of the Roman empire. Had the monks not copied the works of pagan and Christian writers alike, there would have been no Renaissance in the sixteenth century.[51] When stability returned in Europe, these Irish scholars were instrumental in spreading learning, becoming not only the

---

[47] Jim Power, *Eat, drink, be merry-and to hell with guilt, Irish Examiner*,21/12/2007 page 22

[48] Michael Northcott, *Moral Climate: The Ethics of Global Warming* page 273

[49] Sean Mc Donagh, *Climate Change, the Challenge to all of us,* page 176

[50] Ibid, page 184

[51] Thomas Cahill, *How the Irish Saved Civilisation,* Anchor Books, 1996

conservators of civilization, but also the shapers of the medieval mind, putting their unique stamp on Western culture. He says that it is hard to believe that civilisation survived by clinging to rocky outcrops such as Skellig Michael. This time civilisation is threatened, not by barbarians but by climate change. Just as the ruins of ancient Rome became the inspiration for the Renaissance, could the communal sustainable manner in which those Early Christian Irish monks become the inspiration for Christian living in the modern world leading to the renaissance of a civilisation that is in harmony with other life on the planet? The argument against a feeling of helplessness when faced with the magnitude of the problem is not to do nothing but to believe that Ireland can save civilisation, as it has done in the past. This island could become a model for sustainable living and we have already shown that we have the skills and know-how to achieve this. Minister for Energy, Communication and Natural Resources, Eamon Ryan made an announcement in January 2008 pledging funding of €1m towards a world-class, state-of-the-art National Ocean Energy facility in University College Cork, which will have an advanced wave basin for the development and testing of early ocean energy devices. He also pledged a further €2m to support development of a grid-connected wave energy test site near Belmullet, Co Mayo. Electricity generated by wave power will be supplied to the National Grid by 2011. Ireland already has the lead in the development of ocean energy technology. The key is not to condemn countries like China but to show them that there is another means to providing energy security apart from building coal-burning stations or nuclear plants.

There is a small minority of people who deny that global warming is taking place or believe that scientists have got it wrong and that there is no need to seek an alternative way of life. Northcott suggests that these sceptics use the *Pascalian wager.* Act as if climate change is happening and change your lifestyle. 'If global warming is humanly caused, then these actions will turn out to have been essential for human survival and the health of the biosphere. In the unlikely event that it is not, then these good actions promote other goods — ecological responsibility, global justice, care for species - which are also morally right.'[52] We see this in the move to find clean renewable sources of energy are better for the

[52] Michael Northcott, *Moral Climate: The Ethics of Global Warming* page 274

wellbeing of the planet and provide greater energy security in these days of spiralling oil prices. In the past the monastery was the centre for learning. Today the school can once again become the cradle of civilisation, training our young people to build a more sustainable world. Catholic schools can in particular become prophetic leaders in educating students to work towards building a better and more secure planet.

Conclusion

Sean Mc Donagh is clear and unequivocal about what the Catholic Church needs to do in order to meet the challenge of climate change: 'There is a game plan for tackling climate change but it needs to be put into action immediately. In theological terms this is a kairos moment, because the decisions taken by this generation will have huge consequences for future generations. If this generation fails to confront this issue, then no future generation will be able to undo the damage. Every human being and every creature in successive generations will suffer.'[53] He calls for 'more ecological theology in Catholic schools.' He says that one 'cannot presume that Catholics or Catholic missionaries are well informed about the reality and future challenge of climate change unless there is an education campaign at every level of the Catholic Church.' On the practical side he says that religious run institutions should begin the process of reducing their carbon footprint. 'They should set realistic goals and encourage other organizations within the Catholic community to take similar steps.'

In May 2008 the Irish Bishops' published a Pastoral Letter for Catholic education. It says that catholic 'schools seek to form pupils who will find true happiness and strive to give authentic leadership in society. They will do this through the Christian quality of their lives, the unselfish use of their gifts for the common good and their commitment to work for a more just, cohesive and caring human society.[54] It goes on to say that it is fundamental to

---

[53] Sean McDonagh, *Climate change, one of the most serious moral issues of our times*, J/P Alert, July/Aug 2007, http://www.cmsm.org/CMSM_Alert/JulAug07/
[54] *Irish Bishops' Pastoral Letter,Vision 08: A Vision for Catholic Education in Ireland.page 3*
http://www.catholiccommunications.ie/vision08/vision08pastoralletter.pdf

Catholic self-understanding to experience everyday realities as sacramental signs of God working in the world. 'This sacramental view helps pupils to see themselves as the stewards of God's creation and become aware of their ecological responsibility for nature and the environment.'[55]

The problem for Catholic schools is that the Catholic Church has not yet published a pastoral letter on climate change nor have they issued a statement stressing the urgency of the matter for those concerned with education. Everyone, particularly those involved in Catholic education must therefore be open to the views put forward by the WCC and theologians of all denominations who are seeking to devise an adequate ethic of climate change. From a moral point of view we must face up to the truth about global warming and learn to tread more lightly upon the planet. The challenge of climate change demands that action needs to be initiated at school level since no plan of action is coming from outside. It will have to be a *bottom up* rather than a *top down* approach since church leaders and the government appear to have no clearly stated position on the role of schools in tackling climate change. Time is running out and 'if this generation does not act, no future generation will be able to undo the damage that this generation has caused to the planet.' [56] There is a moral obligation for action to be taken immediately in schools to face up to this challenge.

---

[55] *Ibid, page 4.*

[56] *Sean Mc Donagh, The Death Of Life: The Horror Of Extinction, Columba Press, 2005, page 79*

‘Never doubt that a small group of thoughtful committed citizens can change the world; indeed, it is the only thing that ever has.’

Margaret Mead

Chapter 5

Introduction

The debate whether climate change is happening or not is over and this presents a challenge for everyone involved in the field of education. The scientific evidence for climate change is unequivocal. We can no longer deny it is happening - the evidence is all around us. As we have seen, world leaders were slow to react to the mounting scientific evidence. However, there is now a growing momentum at a political level to reduce carbon emissions. The EU aims to take the moral high ground with a proposal to reduce greenhouse gas emissions in the EU by 363m tonnes, or 20%, by 2020. Architect and broadcaster, Duncan Stewart says 'we are talking about a paradigm shift where there is going to be very profound change.'[1] This means that a major challenge lies ahead for school authorities to plan for these changes. Government measures are likely to make it obligatory for schools to reduce their carbon emissions in the years ahead. Even if they do not, we have seen that all schools, particularly Christian schools have a moral obligation to reduce their carbon emissions because of their religious ethos.

As mentioned in the introduction, I have worked for the last three years as a Green-Schools Co-ordinator and I have come to see how the Green-Schools programme is a very effective means of meeting the challenge of climate change in schools. Green-Schools is an international environmental education programme, offering a well-defined, controllable way for schools to take environmental issues from the curriculum and apply them to the day-to-day running of the school. More than twenty thousand schools worldwide are now participating. The Green-Schools programme is an initiative of the Foundation for Environmental Education (FEE), and is known internationally as 'Eco-Schools'. FEE is a non-governmental and non-profit organisation comprising environmental Non-governmental Organisation (NGO) members within countries, established in 1981 to promote sustainable

---

[1] Harry Mc Gee, *Wind of Change*, Irish Times, 26/01/2008, Weekend Review, page1.

development through environmental education. Green-Schools is now in place in almost all European Union Member States, and various countries in Central and Eastern Europe, South Africa and recently introduced in China. An Taisce, the FEE member for Ireland administers the Green-School's programme in Ireland on behalf of FEE. The principle aim of Green-Schools is to make environmental awareness and action an intrinsic part of the life and ethos of a school. This includes the students, teachers, non-teaching staff and parents, as well as the local authority, the media and local business. Green-Schools endeavours to extend learning beyond the classroom and develop responsible attitudes and commitment to caring for the environment both at home and in the wider community. Schools participating in the Green-Schools programme are introduced to environmental themes on a phased basis. Schools are encouraged to focus on one theme in depth, rather than trying to address a wide range of environmental topics in less detail. Only when schools have been awarded the Green Flag are they permitted to introduce a new theme to their respective programmes. The Green Flag award is now established as a well respected and recognised eco-label and is given to schools that complete all the essential elements of the Green-Schools programme. Schools participating in the programme are part of an international movement to bring about sustainable living. The Green-Schools movement is a voluntary program and does not have a statutory footing.

There has been little or no research to date on the effectiveness of the Green-Schools programme in Irish schools. Following the World Summit on Sustainable Development in Johannesburg in 2002 it was recognized that life long learning and education is a driving force towards sustainable development. To promote this, the *UN Decade of Education for Sustainable Development* was established to run from 2005 to 2014 to provide a framework for global action. The goal of the UN Decade is to move society towards sustainable development and the driving force behind this aim is education that is enshrined in its 2006 *Sustainable Development Strategy.*[2] Ireland is obliged to implement the

---

[2] *EU, Sustainable Development Strategy (EU SDS)2, 2006.* http://www.euractiv.com/en/sustainability/sustainable-development-eu-strategy/article-117544 *retrieved 15/10 2007*

strategy and the Department of Education and Science has the overall responsibility for policy development and implementation for education for sustainable development at national level. The partnership agreement, between the government and the social partners[3] commits the government to a review of Ireland's national sustainable development strategy and this began in 2007 and is ongoing at present. ECO-UNESCO carried out a research project on education for sustainable development on behalf of COMHAR which they submitted to this review.[4] Comhar is the forum for national consultation and dialogue on all issues relating to sustainable development. It is one of a number of government advisory bodies. Their report is the most detailed study of education for sustainable development in Irish schools to date. It reviewed education for sustainable development at Primary, Secondary, and at Higher Education Level. It also looked at teacher training. The aim of this research project was to provide an overview of good practice in education and to draw conclusions on the policy frameworks and other mechanisms required to further develop education for sustainable development in Ireland. It found that the Green-Schools programme satisfied all the criteria for education for sustainable development and recommends that all schools should take part in the Green-Schools programme. This chapter examines how schools can best meet the challenge of climate change by promoting and supporting the Green-Schools programme.

The Green-Schools Programme

Once a school is operating the seven step programme, there are no rules or regulations governing how the programme should be implemented in a school. It is left very much to each individual school to adapt the programme to their particular set of circumstances. In 2006 the *www.greenschoolsireland.org* website

---

[3] *Partnership Agreement, Towards 2016,Section 18, page 34 http://www.taoiseach.gov.ie/attached_files/Pdf%20files/Towards2016PartnershipAgreement.pdf retreieved 16/11 2007*

[4] *ECO-UNESCO, Research Project on Education for Sustainable Development in Ireland, Page 5 http://www.comhar-nsdp.ie/ComharDocs/ESD_Research_Project_FINAL_REPORT.pdf. retrieved 21/08/2007*

was set up by An Taisce and since then schools are encouraged to share what has worked for them in implementing the programme and a number of case studies are cited on the site. The programme themes are litter and waste, energy, water, travel and now for the first time, schools in Ireland will be able to move on to address issues associated with climate change, which is inherent in the other themes.

Litter and Waste

Litter & Waste is the first theme that schools undertaking the Green-Schools programme work on. These are major environmental issues in Ireland costing Local Authorities tens of millions of Euros each year cleaning up the environment and disposing of our waste. Figures released by the EPA in January 2008 show that, although the quantity of waste recycled increased between 2005 and 2006, so too did the total quantity of waste generated, resulting in an 8% increase[5] in the amount going to landfill.[6] Until recently most of this waste was landfilled, creating methane gas, one of the most potent greenhouse gasses. O' Mahony and Fitzgerald found students in schools awarded Green Flags are less likely to drop litter and are more likely to participate in local environment projects, conserve water, energy and think about the environment when making a purchase.[7] They found that secondary schools have more waste going to landfill than primary schools. However there is a significant reduction in the amount of waste going to landfill in schools that have adopted the Green-Schools Programme.[8] Recycling levels of glass, paper/cardboard and

---

[5] *EPA, National Waste Report 2006, pages v-viii*
http://www.epa.ie/downloads/pubs/waste/stats/epa_national_waste_report_20061.pdf
retrieved 15/12/2007

[6] *Green-Schools, Litter & Waste*
http://www.greenschoolsireland.org/Index.aspx?Site_ID=1&Item_ID=98
retrieved 15/10/2007

[7] *Dr Michael John O' Mahony and Frances Fitzgerald, The performance of the Irish Green-Schools Programme, 2001, page 30,*
http://www.greenschoolsireland.org/Javascript/tiny_mce/plugins/filemanager_net/files/materials/ireland_research_report_2001.pdf
retrieved 15/12/2007
[8] *Ibid, page 18*

aluminium along with levels of home composting are higher within the homes of Green-Schools students than within the homes of Non-Green-Schools students.[9] Diverting waste from landfill reduces greenhouse gasses and helps to combat climate change. O Mahony and Fitzgerald show that the Green-Schools programme has a positive effect in this regard. Reducing, reusing and recycling waste conserves energy. Recycling one tonne of aluminium cans saves 13,300 kWh of electricity, resulting in a considerable reduction in carbon emissions. The most efficient method of dealing with waste is not to generate it in the first place.

In September 2005, The Green School Committee, St Joseph's Convent Of Mercy Secondary School, Rochfortbridge, made a submission entitled *Our Future In Your Hands* to Westmeath County Council on the proposed Midland Region Waste Management Plan 2005-2010.[10] They called for Green- Schools to be an integral part of the Midland waste management plan because the students of today are the citizens and leaders of tomorrow and if they are educated to control and manage waste the need for landfill and incineration will be greatly reduced in the future. The challenge for all schools is to reduce the amount of waste being generated and in particular, examine policies regarding vending machines and tuck shops which have the effect of generating huge amounts of waste. These have no place in a school truly committed to reducing carbon emissions. Participating in the Green-Schools programme provides the necessary framework and guidelines to do this.

Energy

Schools that have been awarded their first Green Flag move on to work on the energy theme. The aim of the Green-Schools programme in relation to the Energy theme is to increase the awareness of energy issues, especially climate change, among the whole school and wider community and to improve energy efficiency and consumption within the school and the wider

---

[9] *Ibid, page 33*

[10] *'Green School Calls for Greater Support from Authorities as Part of Plan to Reduce Waste; The Local Planet Nov 2005- Feb 2006*

community. Implementing a few simple 'no cost' and 'low cost' ideas for conserving energy dramatically reduces electricity and heating bills within the school and consequentially reduces greenhouse gas emissions and combats climate change. Schools involved in the programme are more aware of energy efficiency and occupant behaviour when it comes to buildings and have been demanding investment in energy efficiency in new buildings and refurbishments. Dr Michael John O' Mahony, An Taisce's Green-Schools Development Officer, speaking in Mullingar in December 2007 to the Midland Green-Schools coordinators spoke of the Department of Education's shortfall in this regard and referred to the ongoing media coverage on poor school fabric.[11] Anecdotal evidence being relayed to him by Green-Schools coordinators around the country is that Boards of Management are finding considerable inertia when they apply to the Department of Education to make the fabric of their schools more energy efficient. Maximum energy efficiency is still not being considered in schools currently under construction or refurbishment. The Government's *National Climate Change Strategy*[12] is committed to a 'programme of installing biomass heating in schools starting with 8 schools in 2007.' Dr O Mahony pointed out that this represents a tiny fraction of the 4000 schools in the country. The strategy states that the government is working with Sustainable Energy Ireland in drawing up plans for schools which will be '2-3 times more energy efficiency than best international normal standards.'[13] This is the limit of the government's commitment to energy efficiency in schools until 2012.

There is a need for all the trustee bodies to form a single lobby group that will act on behalf of all schools to lobby the Department for funds to make all schools under their trusteeship fully energy efficient. In April 2008 representatives from every primary school in Ireland came together for the first time to address an Oireachtas

---

[11] *Dr Michael John O' Mahony, pers. comm..*

[12] *Irish Government, Ireland National Climate Change Strategy 2007-2012 ,page 37*
http://www.environ.ie/en/Environment/Atmosphere/ClimateChange/NationalClimateChangeStrategy/PublicationsDocuments/FileDownLoad,1861,en.pdf
retrieved 15/11/2007
[13] *Ibid, page 37*

Committee on what they say is a mounting funding crisis in schools.[14] Similar cooperation is needed in uniting these representatives with the various bodies representing secondary schools to form an expert group on energy efficiency. This body should then be mandated to lobby the various government departments for energy efficient school buildings. This will be more effective in bringing about change as in this case 'the whole is greater than the sum of the individual parts.' The individual efforts of boards of management are clearly failing to bring about improvements. In the long run investment in energy efficient schools will be cost neutral for the government. Money spent in making schools more energy efficient would otherwise be spent buying carbon credits. A study of 30 'green schools' built with maximum energy efficiency in mind, during the period 2001 to 2006 in ten American states found that these schools use an average of 33 percent less energy and 32 percent less water than conventional schools.[15]In January 2008 Energy Minister, Eamon Ryan announced an initial €8 million grant for schools and hospitals to install energy efficient heating systems.[16] He stated that he wants to promote CHP (combined heat and power) units such as anaerobic digesters which use sewage to create methane gas which is then used to generate heat and electricity. In effect this enables organisations to generate their own heat. He expects the grant to displace the equivalent of 36 million litres of heating oil and reduce carbon emissions by 100,000 tonnes per year. He said he would 'encourage schools and hospitals to generate their own electricity on site using this technology.' The challenge for school authorities is to make full use of these new incentives and plan to gain energy self-sufficiency. The British plan *Building Schools for the Future* designed to rebuild or refurbish all secondary schools in England over 15 years at a cost of £45 billion gives some idea of the level of

---

[14] *RTE, School managers warn of funding crisis, http://www.rte.ie/news/2008/0410/school.html retrieved 10 April 2008.*

[15] *Gregory Kats, Greening America's Schools costs and benefits, 2006, A Capital E Report, page 5 http://www.cap-e.com/ewebeditpro/items/O59F9819.pdf retrieved 15/08/2007*

[16] *Stephen Rogers, Grants for energy-efficient heating systems, Irish Examiner 25/01/2008, page 17*

commitment that is required from the Irish government to achieve maximum energy efficiency in schools.

Water

The third Green-Schools theme is water and this made the headlines in December 2007 with an announcement by the Minister for Education introducing water charges for schools. This will have the effect of making all schools conserve water because after 2009 there will be a charge per unit used. Almost all schools are connected to a public water main. These are administered and maintained by Local Authorities who must comply with the *Water Pollution Acts (1977 and 1990)* and *Urban Waste Water Treatment Regulations* and the *European Water Framework Directive*. In order to comply with these laws, huge amounts of energy go into the treatment of water before it is pumped to consumers. This too, requires a vast amount of electricity. Schools around the country use this treated water for flushing toilets, cleaning purposes etc. In addition, millions of litres are lost through leakage. A tap, dripping once a second wastes about 10,000 litres of water a year.[17] Conserving water has the effect of conserving the energy that goes into its treatment and thereby reduces carbon emissions. The challenge facing schools is to conserve drinking water and invest in rainwater collection for all non drinking water purposes.

Transport

Figures released by the Central Statistics Office in November 2007 show a rise in the number of secondary school children being driven to school with 45% of pupils driven 4km or less.[18] The number of students driving to school doubled from 2,564 in 2002 to 5,131 in 2006. The number travelling by bus was down from 121,000 in 2002 to 108,000 in 2006. The number of secondary school pupils walking to school was also down from 83,000 in

---

[17] *An Taisce, Water, http://www.greenschoolsireland.org/index.aspx?Site_ID=1&Item_ID=234 retrieved 15/12/2007*

[18] *CSO, Travel to Work, School and College, http://www.cso.ie/census/census2006results/volume_12/volume_12.pdf retrieved 21/11/2007*

2002 to 74,000 in 2006. Meanwhile, figures released by the EPA show a 5.2 per cent increase in carbon emmissions from transport in 2006 on 2005 levels.[19] This is the equivalent of almost 680,000 tonnes carbon.Since 1990 there has been a 169% increse in carbon emmissions from the transport sector.[20]Transport now accounts for 20% of the country's greenhouse gasses.

An Taisce ran a Travel Pilot Programme funded by the Dublin Transportation Office (DTO) in partnership with six Local Authorities in twenty nine schools in the Greater Dublin Area (Dublin, Meath, Kildare and Wicklow) from September 2005 to August 2007 to promote sustainable modes of transport on the journey to school.[21] This represented approximately 10,400 children and 545 teachers. As part of their travel action plan, participating schools set their own travel targets, with the ultimate aim of increasing the number of pupils walking, cycling, carpooling or using public transport. Such was its success, funding was approved for the year 2007-2008 in the Greater Dublin Area to extend the project to over 50 schools and it is hoped that it will be rolled out to the rest of the country in the near future. School authorities need to begin promoting sustainable transport modes in order to reduce carbon emissions emanating from transport. The growing trend of students driving cars to school should be discouraged.

Climate change

The topic of Climate Change, and in particular the measuring of carbon dioxide, is not a specific theme for the Green-Schools programme. However working the Litter, Waste, Energy, Water and Travel themes has the net effect of reducing carbon emissions, thereby facing up to the challenge of climate change. Dr Michael

---

[19] *EPA, Ireland's greenhouse gas emissions in 2006, http://www.epa.ie/downloads/pubs/air/airemissions/ghg_provisional_20061.pdf Date released: Jan 15 2008.*

[20] *Ibid*

[21] *An Taisce, Green-Schools Combat Congestion on the School Run!, http://www.greenschoolsireland.org/Index.aspx?Site_ID=1&Item_ID=411. Retrieved 10/10/2007*

John O' Mahony, An Taisce's Green-Schools Development Officer, is currently working on climate change as a theme and he has posted a carbon calculator which has been specially designed for schools on the Green-Schools website.[22] We have seen that climate change can no longer be denied and that there is a moral obligation to do something about it. There is an onus on all schools to reduce their carbon emissions and implementing the Green-Schools programme is an effective means of doing this.

The Green-Schools Programme- Environmental Review

Step two of the Green-Schools Programme involves carrying out an *Environmental Review* to examine the school's environmental impacts in order to identify targets for action and improvement. St Joseph's Secondary School in Rochfortbridge is the first school in Ireland to do this by calculating its ecological footprint. Sonya Quinn, a past pupil of the school and former member of the Green-School committee, currently a lecturer in the University of Limerick worked with the Green-School Committee to achieve this.[23] Ecofootprinting calculates human demand on nature and compares this to nature's ability to regenerate these resources. Currently nature is unable is to meet the demands being placed on it by Western lifestyles. If everyone on the planet was to have a lifestyle similar to the one we have here in Ireland we would need four times the resources that the planet is able to provide at present. In effect our current lifestyle cannot be sustained indefinitely into the future. Reducing one's ecological footprint not only conserves the earth's finite resources, it also reduces carbon emissions. Schools should promote ecological footprinting as part of the Environmental Review.

[22] http://www.greenschoolsireland.org/index.aspx?Site_ID=1&Item_ID=370

retrieved 15/12/2007

[23] *Olga Aughey, A first in Ireland for St. Joseph's, Rochfortbridge, Westmeath Topic, 6th September 2007, page 10*

The Green-Schools Programme – Integration with the Curriculum

The Green-Schools programme requires that environmental studies be integrated into the curriculum (Step5). Climate change and issues related to it are part of the syllabus for science and geography at both junior certificate and leaving certificate level. Also at junior certificate level it is on the syllabus for Civic, Social and Political Education (CSPE). Religion and the Environment is a topic on the leaving certificate Religious Education syllabus and students select a topic of their choice.

However, having environmental issues on the syllabus doesn't always translate into action outside the classroom as a recent study by Quinn and Gaughran shows.[24] They carried out research on students and teachers in second level schools to assess their eco-literacy (understanding of environmental problems) and environmental attitudes with a view to developing strategies for creating a culture of sustainability in Irish schools .They found that the level of eco-literacy was quite low despite the fact that the syllabus has covered environmental issues for a number of years. Almost half of those surveyed said that their single greatest environmental concern was global warming. While most heard about carbon dioxide as a green house gas, a large majority of students had little or no knowledge about the actual greenhouse effect. Most teachers gave adequate or excellent answers revealing that this knowledge is not being passed to students. They also measured the extent to which eco-literacy scores predict attitudes towards the environment and found the more eco-literate a person was, the greater the concern for the environment. They conclude that high levels of literacy does not equate with high levels of eco-literacy. 'The most educated of nations have the higher per capita rates of consumption and leave the deepest footprints.' They note that the way subjects are taught is an important component of cultivating a culture of sustainability. They point out that teachers have not received specialised training or education in the area of sustainability, environmental awareness or eco-literacy, the curricula in the past did not embrace such issues nor did courses in

---

[24] *Quinn, S. and W. Gaughran, Cultivating a Sustainability Culture in Irish Second Level Schools, American Society for Engineering Education. 2007, pages 15-21*

teacher training institutes. A great number of teachers require in-service training to improve their eco-literacy. The challenge now is for the relevant bodies to take note that education for sustainable development should be a key module in the training of teachers, no matter what their subject specialties. In-service training must be provided for all teachers in order for them to integrate education for sustainable development into their subject area. Care must be taken ensure that climate change issues on the syllabus are not lost in translation and that learning in the classroom results in effective action outside the classroom to combat climate change. Above all, teachers must become more aware that climate change issues are not just the forte of the teachers involved in the Green-Schools programme; it is also the responsibility of everyone involved in the education of young people.

The Green-Schools Programme -Informing and involving the school and wider community

The programme requires that the Green-School programme be promoted as much as possible at the level of radio, TV and the press (Step 6). In November 2007 the Green School Committee, St Joseph's Convent Of Mercy Secondary School, Rochfortbridge organised a public lecture in the school entitled 'Climate Change-the Challenge to All of Us.'[25] The head of the Irish Climate Research Unit based at the National University, Maynooth, John Sweeney explained that climate change in Ireland will be largely beneficial as we will have warmer summers with less rainfall and milder winters. Problems such as summer drought and winter flooding will be gradual and can be overcome by planning ahead. Dr Sean McDonagh said that Ireland has one of the world's highest carbon emission rates per capita and while our lifestyles are doing most to cause climate change, we will suffer the least. He said that Christians in a faith based school must halt the world's poorest people being put at even greater risk by climate change. He warned that if this generation fails to confront the issue of climate change, then no future generation would be able to undo the damage. Every human being and every creature in successive generations will suffer. This offers an example of how a school might approach the

---

[25] *Climate Change-A Wake-Up Call for Westmeath, Westmeath Examiner, December 9th 2007*

task of bringing the issue of climate change to the wider community.

The Green-Schools Programme in action

The first of seven steps in the programme is to form a committee. Ideally the committee should be made up of teachers, students and non-teaching staff, preferably the caretaker. The Chairperson is a student and committee meetings should be held at regular intervals. In the case of St. Joseph's, the Transition Years are timetabled for two class periods per week for Green-Schools and together with the Green-Schools Co-ordinator they form the committee. There are a number of advantages in this arrangement. Transition Years are used to working together as a group. They are free from the constraints of a set syllabus and are not under exam pressure. Transition Year by its nature allows students to be very creative and flexible. The problem with this arrangement is that Green-Schools is a whole-school programme and in a school our size with 50 fulltime and part-time teachers and 750 students spread over thirty classes it is difficult to involve everyone. This is true for most secondary schools making it a challenge for the Green-Schools Co-ordinator. Primary schools face other challenges in trying to work the programme.

In 2005 our committee tried holding a Green-Schools meeting once a week at lunchtime attended by a representative from each class. This achieved little for a number of reasons, chief among them were time constraints and the difficulty in getting enough students who were willing to make a commitment. The following year each member of the committee was assigned a class and each week at the same time they visited their class to brief them on Green-School matters. This was made possible by the willingness of all the teachers timetabled at that time to support the programme. This year they have built on that success. The committee carried out the first ever whole-school survey of attitudes towards the environment, which they presented to management in April 2007. This found that 67% of students were most likely to listen to other students when it comes to environmental issues. This was higher among younger students than older students. Students on this year's committee made a series of PowerPoint presentations on various issues relating to climate change such as energy conservation, waste and food miles. During the year, classes in the junior cycle were visited by different teams from Transition Year who instructed them on a wide range of topics using the PowerPoint presentations as teaching aids. Every effort was made to link the PowerPoint presentations with the subject timetabled for the time the class was being visited, an example; the presentation on food miles was given during Home Economics class. Visiting classes was made possible by the co-operation and goodwill of teachers.

The availability of laptops and data projectors is an ongoing challenge facing the committee at present in its attempt to involve the whole school. Most students prepare PowerPoint presentations for class at home but encounter difficulties in school in getting access to IT equipment. Another major difficulty is co-ordinating the timetable for class visits. At present each group is responsible for organizing with the various teachers a time when they can visit classes and inevitably there is a breakdown in communication from time to time. In order for Green-School committees to function efficiently they need to be funded properly and provided with the resources they need such as computer equipment. There is no funding available for Green-Schools committees and the work of a committee is determined to a huge extent by the amount of money raised through fundraising. The challenge for all concerned is to put proper funding in place to enable Green-School committees to make the programme a whole school endeavour.

It can be very difficult to engage students in environmental issues and our experience is that the older the student, the less interested they are in co-operating with the committee. To overcome this problem we initiated a *Green-School Student of the Year* Award. Students who have shown a great commitment in promoting the programme in the school are short listed for this award which is announced at the Annual Prize Giving Ceremony held at the end of the school year and attended by all staff and students. The winner is presented with a trophy and this is covered in the local media.[26] This gives students an incentive to get involved in implementing

---

[26] *Linda O Reilly, Green students emerge top of the class, Westmeath Examiner, Sat 9th June 2007, page16*

the programme. It also recognizes the efforts made by students, who for the most part do most of the practical work.

In our school, the staff is briefed on Green-School matters at staff meetings and also via the staff notice board. Students have also been invited by our Principal to speak at staff meetings. Making it a whole-school programme involves breaking down barriers and opening up new frontiers such as allowing students speak at staff meetings. The whole-school survey referred to above found that all the teachers in our school are willing to work with the Green-School Committee in promoting the programme in their classes.

The understanding and support from school managers is vital in order for a sustainability culture to be fully embraced in schools. True change requires all people involved in the school to work together in the one direction. If only small pockets of people work in isolation towards a goal of carbon reduction, then true change will not happen and any progress they might have been made will certainly fade away. The necessary training and support for Boards of Management needs to be put in place to enable them to make climate change a whole school issue.

The job of Green-Schools co-ordinator is quite intensive and most of the work is done outside of the classroom. At present it is at the discretion of the Principal and Board of Management in each school whether or not it is a Post of Responsibility. Making the Green-School co-ordinator a Post of Responsibility would put it on a statutory footing and require greater accountability on how the job is done. There is unlikely to be any critical review of the programme if the job is done voluntarily. The challenge for everyone at managerial level is to make the position of Green-Schools Co-ordinator a Post of Responsibility so that there will be greater accountability on how the job is done and also remuneration for those doing it.
Green-Schools co-ordinators in the various schools around the country work in isolation from each other. Case studies have been posted on the greenschoolsireland.org website but this is no substitute for proper interaction where co-ordinators can interface with one another and exchange ideas and information. In 2005, I was involved in setting up *Planning Matters* which provides online support for people engaging with the planning system. Our website

has a Forum where people can interact and exchange ideas on a daily basis.[27] I have seen how this can be an invaluable means of connecting people and information. The setting up of a similar forum for Green-Schools co-ordinators is vital to put them in contact with each other so that experience and advice can be shared. The opportunity to network in this way should provide them with the key information they need to overcome the difficulties they experience in implementing the programme.

The Role of Local Authorities

Each of the Local Authorities (City & County Councils) has an Environmental Education Officer (EEO). These officers provide support to schools undertaking the programme. At present over 2,880 schools are participating in the programme in Ireland, representing 67% of all the schools in the country.[28] As the number of schools signing up for the programme grows, so too does the workload for these officers. It is worth noting that Green-Schools is usually only a small part of an officer's role within the local authority. The time has now come for each local authority to appoint a full time Green-Schools Officer. In January 2006, East Cork Area Development (ECAD) initiated a one year pilot programme for schools on Great Island (Cobh), Co. Cork.[29]
A part-time local area Green-Schools co-ordinator was appointed to provide on the ground support to all the schools on Cobh-Great Island participating or wishing to participate in Green-Schools. This pilot project was being funded by East Cork Area Development and the Green-Schools office under the National Development Plan 2000-2006. The role of the co-ordinator involved teacher and student training; provision of relevant local information on environmental management and resources and the development of a best practice network for schools. Most of this work was delivered through seminars, workshops and school visits. This pilot project was very successful in providing support for

[27] http://www.planningmatters.ie/forum/
[28] *An Taisce, Green-Schools in Ireland,* *http://www.greenschoolsireland.org/index.aspx?Site_ID=1&Item_ID=185. Retrieved 20TH June 2008.*
[29] *An Taisce, Green-Schools Cobh-ECAD Pilot Project,* *http://www.greenschoolsireland.org/index.aspx?Site_ID=1&Item_ID=316 Retrieved 20th June 2008*

schools already signed up for the Green-Schools programme and also in encouraging other schools to sign up. The time has now come for the establishment of a similar support mechanism for schools in all local authority areas.

Conclusion

There have been many examples over the years of the hyping of fears: nuclear war, Ebola, avian flu and the Y2K computer crash. In all cases, the threat never fully transpired and the tide of fear ebbed away. Climate change is different. There is nothing ephemeral about global warming. The scientific evidence underlying it presented in Chapter One is unequivocal. Chapter Two demonstrates that the effects of climate change are to be seen all around us. Chapter Three explains that world leaders are beginning to act, albeit very cautiously and are probably not doing enough at present to halt the tide in the short timeframe that is left before climate change is out of control. Chapter Four argues that schools, particularly those that are religious-run are morally bound to face up to the challenge of climate and become a prophetic voice in dealing with it. Chapter Five proposes that the Green-Schools programme is a very effective means of tackling climate change allowing schools to become part of an international force for change that reaches as far as China. Dr Michael John O Mahony in his role as An Taisce's Green-Schools Development Officer has calculated an average 7,000-10,000 tonnes per annum diversion of waste from landfill from schools that implemented the programme, 10-15% increase in recycling and composting levels in homes of Green-Schools students and a reduction in energy use and consumption by green-schools of around 20-30% leading to7,000-10,000 tonnes reduction in CO2 per annum.[30] Green-Schools has a proven track record in creating a more sustainable culture but if it is to tackle the issue of climate change it needs to be supported and funded properly. Only time will tell whether or not we took the issue of climate change seriously enough to heed this call to face up to the challenges that it poses.

---

[30] *Pers. Comm.*

'Our generation has inherited an incredibly beautiful world from our parents and them from their parents. It is in our hands whether our children and their children inherit the same world.'

Richard Branson

Bibliography

Arrhenius, Svante 'On the Influence of Carbonic Acid in the Air Upon the Temperature of the Ground', *Philosophical Magazine* 41: (1896) 237-76.

Attenborough, David, 'Climate Change is the Major Problem Facing the World', *Independent,* UK, May 24, 2006

Berry, Thomas, *Dream of the Earth*, Sierra Club Books, 1988

Caldwell, Simon , 'The Pope condemns the climate change prophets of doom', *Daily Mail,* 14th November 2007

Callendar, G.S. 'The Artificial Production of Carbon Dioxide and Its Influence on Climate', *Quarterly J. Royal Meteorological Society* **64**: (1938). 223-40.

Cahill, Ann , 'Hot air; report criticises Irish plan to buy carbon credits', *Irish Examiner,* 13th June 2007

Cahill, Ann, 'Ireland must reduce its CO2 emissions by a fifth by 2020', *Irish Examiner,* 24 January 2008

Cahill, Thomas, *How The Irish Saved Civilization*, Hodder & Stoughton, 1995

Chan, Margaret, *Climate change will erode foundations of health,* http://www.who.int/mediacentre/news/releases/2008/pr11/en/index.html, 7th April 2008.

Collins Dan, Global warming fears for south and east, Irish Examiner, 10 June 2008, page 1

Connor, Steve , 'Climate change causes new epoch', *London Independent*, 26 January 2008

Daly, Cardinal Cahal, *The Minding of Planet Earth*, Veritas, 2004

Deane-Drummond, Celia, *The Ethics of Nature*, Blackwell Publishing, 2004

Edwards, Denis, *Jesus the Wisdom of God*, Orbis, 1995

Edwards, Denis, *Ecology at the Heart of Faith*, Orbis, 2006

Fell, Charlie , 'Seeds of Change, food prices could easily double in the next seven years', *Irish Times*, 8th June 2007

Gaffney, Owen , 'Carbon trading: A new global commodity'. *Irish Examiner,* 7th Jan 2008

Hood, Marlowe, 'Failure to tackle global warming criminal', *Irish Examiner*, 11th Nov 2007

Huntley, Brian et al, *A Climatic Atlas of European Breeding Birds*, Lynx Edicions, 2008

Keeling, Charles D. (1978). 'The Influence of Mauna Loa Observatory on the Development of Atmospheric CO2 Research.' In *Mauna Loa Observatory. A 20th Anniversary Report. (National Oceanic and Atmospheric Administration Special Report , September 1978)*, edited by John Miller, pp. 36-54. Boulder, CO: NOAA Environmental Research Laboratories

Kerr, Aine , 'A fifth of plant life faces climate change risk', *Irish Examiner* 30th November 2007

Lederer, Edith, 'Production of biofuels is a crime', *London Independent*, 27th Oct. 2007

Mac Connell, Seán, 'Ireland's iconic native birds at risk from climate change', *Irish Times,* 16th January 2008

Mann, Michael E., et al. 'Northern Hemisphere Temperatures During the Past Millennium: Inferences, Uncertainties, and Limitations.' *Geophysical Research Letters* **26**: (1999), pp. 759-62.

Mc Donagh, Sean, *The Greening of the Church*, Continuum International Publishing, 1990

Mc Donagh, Sean, *Dying for Water*, Veritas, 2003

Mc Donagh, Sean, *The Death of Life-The Horror of Extinction*, Columba Press, 2004

Mc Donagh, Sean, *Climate Change-The Challenge to All of Us*, Columba Press, 2006

Mc Fague, Sally, *Life Abundant-Rethinking Theology and Economy for a Planet in Peril*,
Fortress Press, Minneapolis, 2001

Mc Gee, Harry, 'Wind of Change', *Irish Times*, 26/01/2008

Mc Greevy, Ronan, 'Warmest year since the 1880s', *Irish Times,* 31st December 2007

McGreevy, Ronan, 'Climate change outpacing science', *Irish Times*, 21st November 2007

Montgomery, Hugh , 'Prognosis looks increasingly grim for the health of nations-Threats to the water supply, crop failure and mass migration are all looming', *Times* London, 10 th March 2007

Northcott, Michael, *A Moral Climate: The Ethics of Global Warming*, Darton, Longman & Todd Ltd., 2007

O' Mahony, Michael J., and Fitzgerald, Frances, *The performance of the Irish Green-Schools Programme*, An Taisce 2001

O Reilly, Linda, 'Green students emerge top of the class', *Westmeath Examiner*, 9th June 2007

Oreskes Naomi, 'Beyond the Ivory Tower : The Scientific Consensus on Climate Change,' *Science* 3 December 2004: Vol. 306. No. 5702, p. 1686

Power, Jim, 'Eat, drink, be merry-and to hell with guilt', *Irish Examiner*, 21st December 2007

Quinn, S. and W. Gaughran, 'Cultivating a Sustainability Culture in Irish Second Level Schools,' *American Society for Engineering Education*. 2007

Rasmussen, Larry, *Earth Community Earth Ethics*, Orbis, 1996

Rogers, Stephen , 'Grants for energy-efficient heating systems', *Irish Examiner* 25th January 2008

Sample, Ian, 'Acidic seas may kill 98% of world's reefs by 2050', *The Guardian*, 14th December 2007

Sample, Ian, 'Warming hits "tipping point"-Climate change alarm as Siberian permafrost melts for first time since ice age', *The Guardian*, 19th August 2005

Staunton, Denis, *Food prices could lead to starvation, say IMF and World Bank*, Irish Times 14th April 2008,

Suess, Hans E. 'Natural Radiocarbon and the Rate of Exchange of Carbon Dioxide between the Atmosphere and the Sea.' In *Nuclear Processes in Geologic Settings*, edited by National Research Council Committee on Nuclear Science, (1953), Washington, D. C.: National Academy of Sciences. pp. 52-56.

Sweeney, J. *Climate Change: Indicators for Ireland*, Environmental Protection Agency, Johnstown Castle, Wexford, 2002

Sweeney, J. and Fealy, R. 'A Preliminary Investigation of Future Climate Scenarios for Ireland, Biology and Environment', *Proceedings of the Royal Irish Academy*, 102B, (2003) pp. 121-128.

Sweeney, J. and McElwain, L. 'Climate Change in Ireland: Recent changes in temperature and precipitation', *Irish Geography*, 36(2), (2003) pp.97-111.

Sweeney, J. and McElwain, L. *Key Meteorological Indicators of Climate Change in Ireland,* Environmental Protection Agency, Johnstown Castle, Wexford, 2007

Vidal, John , 'Global food crisis looms as climate change and fuel shortages bite', *The Guardian*, 3rd Nov. 2007

List of Websites

http://www.catholiccommunications.ie/Pressrel/pressrelease-archive2007.html

http://www.cmsm.org/CMSM_Alert/JulAug07/
http://environment.independent.co.uk/article570935.ece

http://www.g-8.de/Content/DE/Artikel/G8Gipfel/Anlage/2007-06-07-gipfeldokument-wirtschaft-eng,property=publicationFile.pdf

http://www.greenpeace.org.uk

http://www.guardian.co.uk

http://www.hm-treasury.gov.uk

http://unfccc.int/kyoto_protocol/items/2830.php

http://www.ipcc.ch/

http://unfccc.int/meetings/cop_13/items/4049.php

http://nobelpeaceprize.org/eng_lect_2007c.html

http://www.oikoumene.org/en/

http://www.unep.org

http://www.wcc-coe.org/wcc/what/jpc/cop6-e.html

http://www.who.int/globalchange/climate/summary/en/index.html

http://www.vatican.va/holy_father/benedict_xvi/messages/peace/documents/hf_ben-xvi_mes_20071208_xli-world-day-peace_en.html

www.bbc.co.uk

www.ceist.ie

www.environ.ie

www.epa.ie

www.greenschoolsireland.org

www.greenschoolsireland.com

www.planningmatters.ie

www.sciencemag.org/cgi/content/full/306/5702/1686